D0848957

MEMORY ISSUES IN EMBEDDED SYSTEMS-ON-CHIP

Optimizations and Exploration

MEMORY ISSUES IN EMBEDDED SYSTEMS-ON-CHIP

Optimizations and Exploration

by

Preeti Ranjan Panda
Synopsys, Inc.

Nikil Dutt
University of California/Irvine

Alexandru Nicolau
University of California/Irvine

KLUWER ACADEMIC PUBLISHERS
Boston / Dordrecht / London

Distributors for North, Central and South America:
Kluwer Academic Publishers
101 Philip Drive
Assinippi Park
Norwell, Massachusetts 02061 USA
Telephone (781) 871-6600
Fax (781) 871-6528
E-Mail <kluwer@wkap.com>

Distributors for all other countries:
Kluwer Academic Publishers Group
Distribution Centre
Post Office Box 322
3300 AH Dordrecht, THE NETHERLANDS
Telephone 31 78 6392 392
Fax 31 78 6546 474
E-Mail <orderdept@wkap.nl>

Electronic Services <http://www.wkap.nl>

Library of Congress Cataloging-in-Publication Data

A C.I.P. Catalogue record for this book is available
from the Library of Congress.

Printed on acid-free paper.

Printed in the United States of America

Contents

List of Figures

List of Tables

Preface

As embedded systems realize the convergence of computers, communications, and multimedia into consumer products, system designers will need rapid exploration tools to evaluate candidate architectures for Systems-on-a-Chip (SOC) solutions. Such exploration tools will facilitate the introduction of new products customized for the market under increasingly shorter time frames.

Traditionally, the memory subsystem has been a major bottleneck in the design of high-performance processor-based systems; due to the rapidly increasing gap between memory and processor performance, this memory subsystem bottleneck becomes even more critical for memory-intensive embedded system applications that process large volumes of data under demanding throughput, cost and power constraints. This book covers techniques for optimization of system-level memory requirements, and exploration of candidate memory architectures for implementing processor-core-based embedded systems.

Modern design libraries frequently consist of predesigned Intellectual Property (IP) blocks such as embedded microprocessor cores and memories. The book outlines techniques for efficient utilization of memory-based modules such as Data Cache, Scratch-Pad Memory, and DRAM during design exploration and system synthesis. The behavioral specification of such embedded systems can be realized in hardware and software. In the case of memory accesses mapped into hardware (i.e., high-level synthesis), models are presented for incorporating off-chip memory access modes, and optimization techniques are proposed for exploiting efficient access modes of modern DRAMs in high-level synthesis. Additionally, techniques are presented for storing data in memory with the objective of enhancing performance and power characteristics of embedded systems.

Software realization of the system behavior requires mapping into an embedded processor, and is characterized by several unique optimization opportunities that have not been addressed traditionally in general-purpose compiler and architecture research.

Applications involving large arrays typically store such data in off-chip DRAMs, with a small amount of on-chip memory, implemented as caches and scratch-pad memory. A technique is presented for memory data organization in order to improve cache performance. A technique for partitioning of data between on-chip scratch-pad memory and off-chip memory with the objective of minimizing off-chip memory accesses is also presented.

In the context of architectural exploration, an analytical technique is presented for rapid evaluation of the impact of different on-chip memory configurations on the expected performance of an application. This analytical exploration scheme considers the appropriate sizing and parameterization of data cache and scratch-pad memory, and provides important feedback to the system designer, who can then select an appropriate memory architecture for the embedded system.

The book concludes with a case study of memory system exploration for the MPEG decoder standard, where the utility of the memory optimizations and exploration scheme is demonstrated through a number of experiments that yield candidate memory architectures with different size-performance characteristics. The experiments show that the analytical estimations used in the exploration provide good approximations to the actual simulations, thereby providing valuable feedback to the system designer in the early phases of system design. This case study using an exploration approach should prove to be directly usable by practitioners involved in the design of embedded systems-on-chip.

Audience

This book is designed for different groups in the embedded systems-on-chip arena.

First, it is designed for researchers and graduate students who wish to understand the research issues involved in memory system optimization and exploration for embedded systems-on-chip. The issues addressed in this book touch upon the traditionally disjoint areas of computer architecture, compiler design, and computer-aided design, and also describe the emerging trend of reusing system-level IP blocks such as processor cores and memories.

Second, it is intended for designers of embedded systems who are migrating from a traditional micro-controller centered, board-based design methodology to newer design methodologies using IP blocks for processor-core-based embedded systems-on-chip.

Third, this book can be used by CAD tool developers who wish to expand their application base from a hardware synthesis target to embedded systems that combine significant amounts of software with hardware. CAD tool developers will be able to review basic concepts in memory hierarchies and expand the reach of their optimization techniques to cover compiler and architectural aspects of embedded systems.

Finally, since this book illustrates a methodology for optimizing and exploring the memory configuration of embedded systems-on-chip, it is intended for managers and system designers who may be interested in the emerging capabilities of embedded systems-on-chip design methodologies for memory-intensive applications.

PREETI RANJAN PANDA, NIKIL DUTT, ALEXANDRU NICOLAU

Acknowledgments

We would like to thank all of our colleagues, students and visitors at the Center for Embedded Computer Systems (CECS) at the University of California, Irvine, who have, over the years, provided an intellectually stimulating environment for pursuing research in this area. The weekly research meetings on Friday afternoons provided a fertile ground for testing our ideas, as well as for discovering new and interesting research problems. Discussions with our colleagues, including Professors Daniel Gajski, Rajesh Gupta, Daniel Hirschberg, and Tomas Lang, helped shape the direction of some chapters in the book. Additionally, discussions with the following visitors to the CECS helped give a broader perspective for this work: Professors Gianfranco Bilardi, Francky Catthoor, Rolf Ernst, Hi-Seok Kim, Peter Marwedel, Hiroshi Nakamura, Wolfgang Rosenstiel, Allen Wu, Hiroto Yasuura, and Drs. Florin Balasa, Vijay Nagasamy, and Takao Onoye.

We had a very fruitful collaboration with Professor Hiroshi Nakamura of the University of Tokyo, during his visit to our laboratory in 1996. He made several original contributions to the Data Alignment Technique described in Chapter 4 and we thank him for his insight, input and friendship.

Asheesh Khare deserves special thanks for finding several typos and minor mistakes in a previous version of this manuscript.

This work was partially supported through grants from the National Science Foundation (CDA-9422095 and MIP-9708067), Semiconductor Research Corporation (95-DJ-146), and Office of Naval Research (N00014-93-1-1348). The authors are grateful for their support.

Finally, we thank our respective families and friends for their support during the writing of this book.

1 INTRODUCTION

1.1 EMBEDDED SYSTEMS-ON-CHIP DESIGN

The design of modern digital systems is influenced by several technological and market trends, including the ability to manufacture ever more complex chips, but under increasingly shorter time-to-market. This particular combination of increasing design complexity coupled with shrinking product design cycle times has fueled the need for design reuse in the IC-design industry. System-level design reuse is enabled by modern design libraries, which frequently consist of pre-designed megacells such as processor core, memories, co-processors, and modules implementing industry-standard functions. Systems-on-Chip (SOC) and Systems-on-Silicon (SOS) are terms used to describe products combining and integrating such megacells onto a single silicon chip; this trend is expected to continue with shrinking feature sizes, and with the capability of embedding DRAMs with logic on the same chip.

Market pressures have also necessitated the need for increasing product differentiation and specialization. Consequently, modern digital circuits can be broadly classified into two categories: General-purpose systems, which can be used in a wide range of applications, e.g., microprocessors, microcontrollers, and memories; and Application-specific systems or Embedded systems, which perform the function of a single or dedicated application, e.g., chips designed for fax/modem and automotive engine control. Time-to-market considerations have driven the need for automation in the embedded systems design industry. Although a completely manual approach relying on an expert designer's intuition, helps reduce chip area and improve performance and power metrics, the time spent by a human designer to optimize the complex chips designed today is prohibitively large, and the manual approach becomes impractical. Thus, most application-specific systems rely on the use of Computer Aided Design (CAD) tools that generate the electronic circuitry from a high-level description of the System-on-Chip.

If we focus on the digital parts of such embedded SOC, a key feature is the coexistence of *programmable processor cores*, *custom designed logic*, and *different types of memory* on a single chip. Whereas the design of these subsystems were traditionally done by separate teams of designers with specialized skills in the domains of computer architecture (for processors and memory hierarchies), compiler design (for the embedded software), and digital CAD (for the custom logic), the design of embedded SOC will require a close coupling and a good knowledge of all these domains.

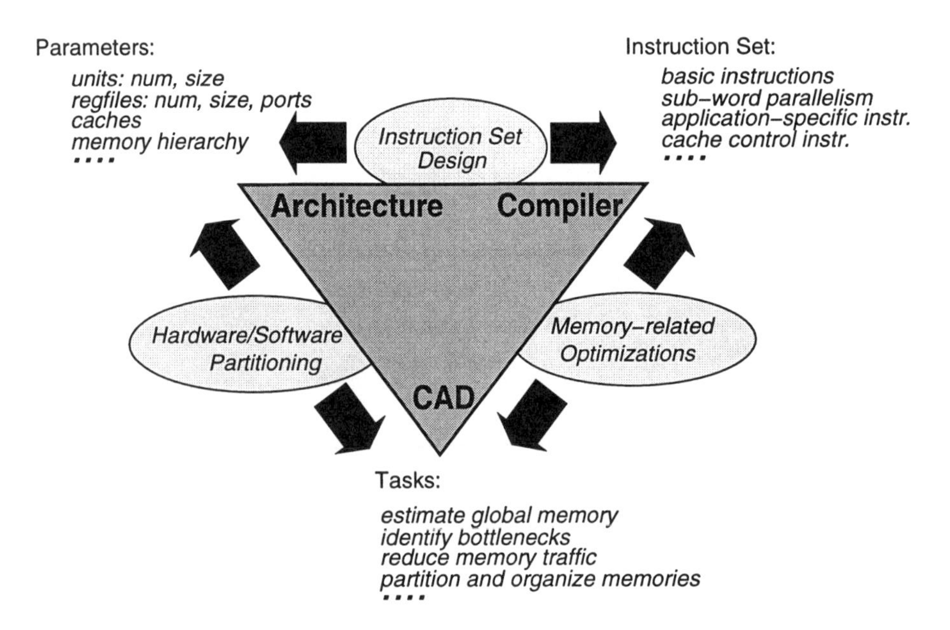

Figure 1.1. Compiler-Architecture-CAD Coupling for Memory-Intensive Embedded Systems

This coupling is illustrated for memory-intensive embedded applications in Figure 1.1. The architecture of modern-day programmable processor core-based systems have several architectural features that are parameterizable, including functional units (number, bit-widths), cache (size, organization), memory (number and size of register files, number of special purpose registers), and even the interconnections within the processor data paths. The instruction sets of these processor cores can also be customized; for memory-intensive applications, the processor core's basic instruction set can be augmented by specialized instructions such as those exploiting subword data (SIMD) parallelism, and cache control instructions. Such processor core customization is typically done by profiling target applications and tuning the processor architecture and its instruction set for meeting demanding performance and power requirements. Customized processors, in turn, need sophisticated compiler technology to generate efficient code for different versions of the processors. Hence, the use of parameterized processor cores requires a tight coupling and a good knowledge of *computer architecture* and *compiler design*. Furthermore, the task of hardware/software partitioning is required to migrate slower system functionality to co-processors and custom logic based on performance bottlenecks. This partitioning step requires a close coupling between *computer architects* and *CAD/hardware designers*. Finally, as memory requirements for embedded applications increase (particularly in the emerging areas of multimedia products), system designers need to analyze global memory requirements, identify bottlenecks, reduce memory traffic for improving speed and/or reducing power, and balance memory utilization between the processors, co-processors and the custom logic on the chip. This requires a close coupling between *hardware CAD experts* and *compiler experts*. Thus, design challenges for embedded SOC require a good knowledge of, and a close coupling between computer architecture, compilers, and electronic CAD.

1.2 SYNTHESIS, COMPILATION AND ARCHITECTURE FOR EMBEDDED SOC

The process of automatic transformation of a more abstract behavioral description into a less abstract (hence, more detailed) level of specification is called *Synthesis*. The existence of a multitude of implementation possibilities for a single behavior makes the search for an optimal design a challenging task for synthesis tools, which operate at various levels of abstraction. *Layout Synthesis* is the task of generating mask-level layouts from logic gate-level descriptions. *Logic Synthesis* is the task of generating optimized logic circuits from register-transfer-level descriptions. *Behavioral Synthesis* or *High-level Synthesis* (HLS) is the task of generating a register-transfer-level description from a behavioral description. Finally, *System Synthesis* refers to the task of automatic generation of more complex systems which could have a hybrid implementation, consisting of both hardware and software components.

The increase in level of abstraction is accompanied by a corresponding increase in complexity of the building blocks that constitute the design library available to the synthesis tools. For example, logic synthesis tools construct an implementation consisting of a network of logic gates. Behavioral synthesis tools realize an implementation in terms of register-transfer components such as adders, multipliers, and multiplexers. In the case of system synthesis, the design library frequently consists of complex blocks such as microprocessor cores, memories, numeric co-processors, and modules implementing standardized functions such as JPEG[Wal91]. For example, the CW33000 processor core from LSI Logic, the TMS320 series of DSP processors from Texas Instruments, the StrongARM processor from Advanced RISC Machines are available in the form of *embedded processor cores*, in addition to the standard packaged form. Memory-based mega-modules are widely used in the industry today in the form of cache, *Scratch-Pad SRAM*[LSI92] and embedded DRAM[Wil97, Mar97].

Embedded processor-based system design is characterized by various optimization opportunities that have not been addressed by traditional compilation and architecture research for general purpose processors. Traditionally, compiler research has focussed on techniques for efficient generation of code for a given architecture. While embedded systems do benefit from these techniques, embedded SOC also present an additional degree of freedom arising from the flexibility of the underlying architecture itself. This leads to various novel research directions involving both compilation and architectural issues, such as application-specific instruction set design, and application-specific memory architectures. Such issues have not been addressed in traditional architecture research, because the designer of a general-purpose microprocessor has to optimize the architecture to execute a large variety of applications efficiently, and consequently, usually selects the features that lead to the best average performance over a set of benchmark applications.

In this book, we focus on the topic of memory-related optimizations for embedded systems, and present several instances of optimizations in the Compiler-Architecture-CAD space of Figure 1.1. We also present an exploration strategy for determining an efficient on-chip data memory architecture for embedded SOC, based on an analysis of a given application.

1.3 EMBEDDED SYSTEM ARCHITECTURE

Figure 1.2 shows a sample design flow and architectural model of an embedded processor core-based system, consisting of a processor core, on-chip memory, and custom synthesized block, interfacing with off-chip memory (e.g., DRAM). The system's behavioral specification is first analyzed in a Hardware/Software partitioning step [GVNG94, CGL96] to determine what part is synthesized into hardware and what part is implemented as software. The custom synthesized hardware block in Figure 1.2 is often generated by high-level synthesis [GDLW92], and performs the functions specific to the application that are mapped to hardware, possibly for per-

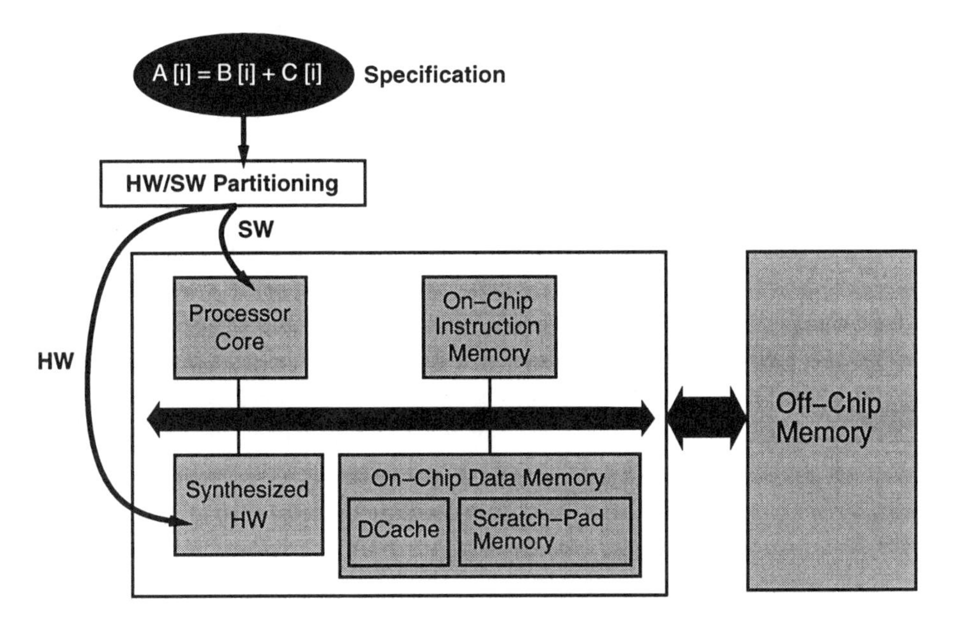

Figure 1.2. Embedded Processor-based Synthesis Flow

formance considerations. The portion of the specification mapped into software is executed on the *Processor Core* block. In addition, the architecture has to support memory devices for storing the code executed by the processor core and for storing the data used during computation.

The on-chip memory can be implemented as a combination of cache, ROM, *Scratch-pad SRAM*, and more recently, as embedded DRAM. Instruction and data cache are fast, on-chip memory forming an interface between the processor and the off-chip DRAM, reducing the effective memory access time by storing recently accessed instructions and data [HP94]. The on-chip instruction memory in Figure 1.2 consists of instruction ROM and/or instruction cache.

The on-chip data memory in Figure 1.2 consists of data cache and Scratch-Pad memory. Scratch-Pad memory is on-chip data memory, to which the assignment of data is compiler-controlled – a portion of the total data memory space is stored in on-chip SRAM, instead of being accessed through the cache. The Scratch-Pad memory is used to store critical data, so that the data is guaranteed to be present on-chip during every access, and is never subject to cache misses. Current embedded SOC, particularly in the area of graphics controllers and multimedia chips, have Scratch-pad memory in the form of *embedded DRAM*[Wil97, Mar97]; this trend is expected to continue as logic is merged with DRAM in future SOC. Although we associate Scratch-Pad memory with only data in this book, it could also be useful in the context of instructions. Usually, the code executed by the embedded processor is small enough to reside on-chip, but if the code is too large and has to be stored off-chip, Scratch-Pad memory could be utilized to store frequently executed instruction segments.

The synthesized hardware block uses the same bus to interface with the memory as the processor core. However, while the processor core requires a memory controller (not shown in diagram) to interface with off-chip memory, the synthesized hardware could integrate the function of the memory controller (or parts of it) within itself.

In addition to the blocks shown in Figure 1.2, an embedded system may contain other modules such as co-processors. The processor core (and co-processor, if any) can be assumed to be fixed for our purpose, although researchers are also studying the impact of application-specific tailoring of the processor core itself [LMP94, SYO+97]. The synthesized hardware block is the result of high-level synthesis of the portion of the specification mapped into hardware. The on-chip instruction and data memory can be tailored for the specific application; clearly, the optimal on-chip memory configuration is a function of the characteristics of the application being synthesized. In this context, the interface of both the processor core and synthesized hardware block with on-chip and off-chip memory provides a rich set of interesting optimization opportunities with respect to performance and power, that form the core of this book.

1.4 ORGANIZATION OF THE BOOK

In this book, we address optimization and exploration issues that arise in different subsets of the embedded system architecture shown in Figure 1.2. Behavioral memory accesses could be realized in either software or hardware – we assume that this decision has been made in a previous hardware/software partitioning step, or has been done manually by the designer. In the case of memory accesses mapped into hardware (i.e., high-level synthesis), we address the interface between the synthesized hardware block and off-chip memory. In the case of memory accesses mapped into software, we address a subset of the architecture in Figure 1.2, with the processor core, the on-chip data memory blocks, and off-chip memory. The processor core/instruction cache interface has been addressed by researchers [TY96a, TY96b, KLPMS97]. The problem of an integrated optimization and exploration for the entire architecture shown in Figure 1.2 is a challenging research topic; we address several aspects of this problem in this book. Specifically, we make the following contributions:

- We present models for incorporating off-chip memory access modes in behavioral synthesis, and optimization techniques for exploiting efficient memory access modes.

- We present techniques for storing data in memory with the objective of enhancing performance and power characteristics of embedded systems.

- We present a fast exploration strategy for evaluating the effect of different on-chip memory architectures on system performance. This exploration provides valuable feedback to the system designer, who can then select an optimal memory architecture for his application. We focus on data memory exploration.

The rest of this book is organized as follows:

Chapter 2: Background. We review concepts in memory hierarchies, and discuss previous and related work for memory issues in embedded systems.

Chapter 3: Off-chip Memory Access Optimizations. We describe the organization of off-chip memories, identifying optimization possibilities for performance and power. We then present techniques for incorporating off-chip memory organization during synthesis to enhance performance, and for storing application data in off-chip memories to minimize power consumption.

Chapter 4: Data Organization: The Processor Core/Cache Interface. We present techniques for aligning data in off-chip memory targeting improvement in cache performance, and hence, reduction in memory traffic.

Chapter 5: On-chip Vs. Off-chip Memory: Utilizing Scratch-Pad Memory. We present an algorithm to efficiently partition application data between on-chip Scratch-Pad memory and off-chip memory so as to maximize performance.

Chapter 6: Memory Architecture Exploration. We present an analytical scheme for rapidly evaluating the impact of different on-chip memory configurations on the expected performance of an application.

Chapter 7: Case Study: MPEG Decoder. We present a case study of our data memory architecture exploration strategy on the MPEG decoder application.

Chapter 8: Conclusions. We summarize the work presented in the book, identify extensions to this work, and conclude with possible directions for future research.

2 BACKGROUND

2.1 INTRODUCTION

The design of the memory subsystem has received considerable attention among computer designers because of the crucial role played by memory in processor-based computation: the execution of every single instruction involves one or more accesses to memory. With the explosion of the memory requirements of application programs and the outpacing of the memory access speeds by processor speeds, the memory interface quickly became a major bottleneck. Since a larger memory implies a slower access time to a given memory element, computer architects have devised a *memory hierarchy*, consisting of several levels of memory, where higher levels comprise larger memory capacity and hence, longer access times. The memory hierarchy operates on the principle of *locality of reference*: *Programs tend to reuse instruction and data they have used recently* [HP94]. Thus, the first time an instruction or data is accessed, it might have to be fetched from a higher memory level, incurring a relatively higher memory access time penalty. However, it can now be stored in a lower memory level, leading to faster retrieval on subsequent accesses to the same instruction or data.

The different memory levels used in most processor architectures are usually: register, cache memory, main memory, and secondary memory.

Registers. Registers afford the fastest access to the Central Processing Unit (CPU), and are usually few in number (typically less than 32) in most architectures.

Cache Memory. Cache memory is the next memory level and stores recently accessed memory locations – *instruction cache* stores recently accessed instructions and *data cache* stores recently accessed data. The two are sometimes combined into a single cache. In most recent machines, cache memory usually resides in the same chip as the CPU. Access times for cache memory usually range from one to a few CPU cycles. Caches in modern commercial general-purpose microprocessors could be as large 64 KB or 128 KB.

Main Memory. Often, the entire memory space required for program execution could exceed the main memory size. Hence, the *virtual memory* [HP94] space for the program is divided into pages, with the recently accessed pages residing in main memory. Thus, main memory, in principle, is analogous to cache memory. Access to the main memory is relatively slow – in the range 5 to 50 CPU cycles for modern architectures. Main memory sizes for modern processors are typically upwards of 32 MB.

Secondary Memory. This is the highest level of the memory hierarchy. The entire virtual memory space required for an application can be stored in secondary memory, which is usually mapped to the disk on a computer. Access to secondary memory is the slowest (hundreds of CPU cycles), and the memory capacity is the largest (several gigabytes of storage).

Figure 2.1 shows a generalized memory hierarchy with the levels of hierarchy described above. The register file is usually incorporated into the CPU. The cache, in turn, could consist of multiple levels of hierarchy, of which the lower levels are usually located on-chip with the processor, and higher levels could be in off-chip SRAM (Static Random Access Memory). The main memory is typically implemented in DRAM (Dynamic Random Access Memory) technology, which affords higher density than SRAM, but lower access speed.

2.2 EXPLOITING MEMORY HIERARCHY

A variety of techniques have been adopted for exploiting the general memory hierarchy structure of Figure 2.1 with the goal of improving program performance. The techniques can be broadly categorized into hardware techniques and software techniques. We limit our discussion to cache memory alone.

Cache misses occur when instructions or data requested by the processor are not present in the cache, and need to be fetched from main memory. Cache misses can be classified into three categories [HP94]:

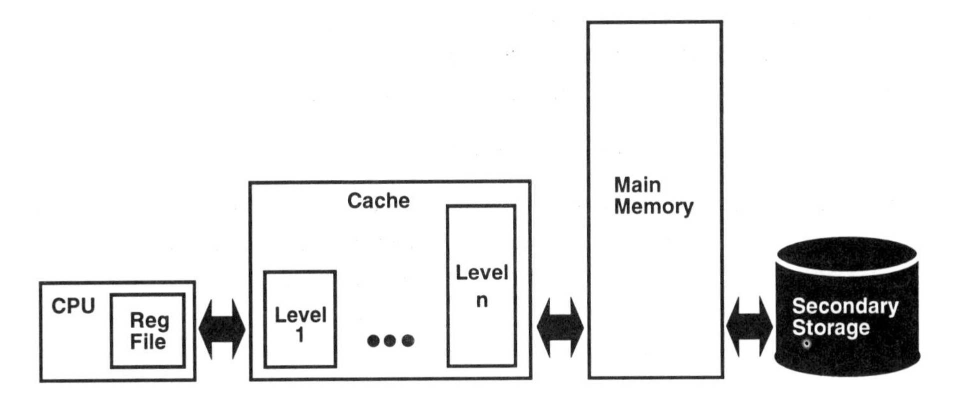

Figure 2.1. Hierarchical Memory Structure

Compulsory misses – caused when a memory word is accessed for the first time.

Capacity misses – caused when cache data that would be needed in the future is displaced due to the working data set being larger in size than the cache.

Conflict misses – caused when cache data is replaced by other data, in spite of the presence of usable cache space.

The *Cache Miss Ratio* is defined by the equation:

$$\text{Cache Miss Ratio} = \frac{\text{Number of Cache Misses}}{\text{Number of Memory Accesses}} \qquad (2.1)$$

Cache Hit Ratio is defined as : 1 − Cache Miss Ratio.

2.2.1 Hardware Techniques

Hardware techniques for exploiting memory hierarchy involve architectural improvements to cache design. The most important technique in this category – *cache lines* and *cache associativity* have become standard cache parameters.

2.2.1.1 Cache Lines. A cache line consists of a set of memory words that are transferred between the cache and main memory on a cache miss. A longer cache line reduces the compulsory misses, but increases the *cache miss penalty* (the number of CPU cycles required to fetch a cache line from main memory), and would also increase the number of conflict misses.

2.2.1.2 Cache Associativity. Associativity is defined as the number of different locations in the cache that can be occupied by a memory element. A *direct-mapped* cache has associativity = 1, implying that any memory element can be mapped to exactly one location in the cache. If the main memory is divided into blocks of the cache line size, the mapping function is given by the equation:

$$\text{Cache Line} = (\textit{Block Address}) \bmod (\textit{Cache Size}) \tag{2.2}$$

where *Block Address* refers to the main memory block number and cache size is the number of lines in the cache. In Figure 2.2(a), element A, which is located in main memory block number 33, maps to cache line number $33 \bmod 8 = 1$ in the direct-mapped cache of Figure 2.2(b). Usually, the number of cache lines is an exact power of two (of the form 2^k), so that the mapping function is very simple to implement – the k lower order bits of the address gives the cache line.

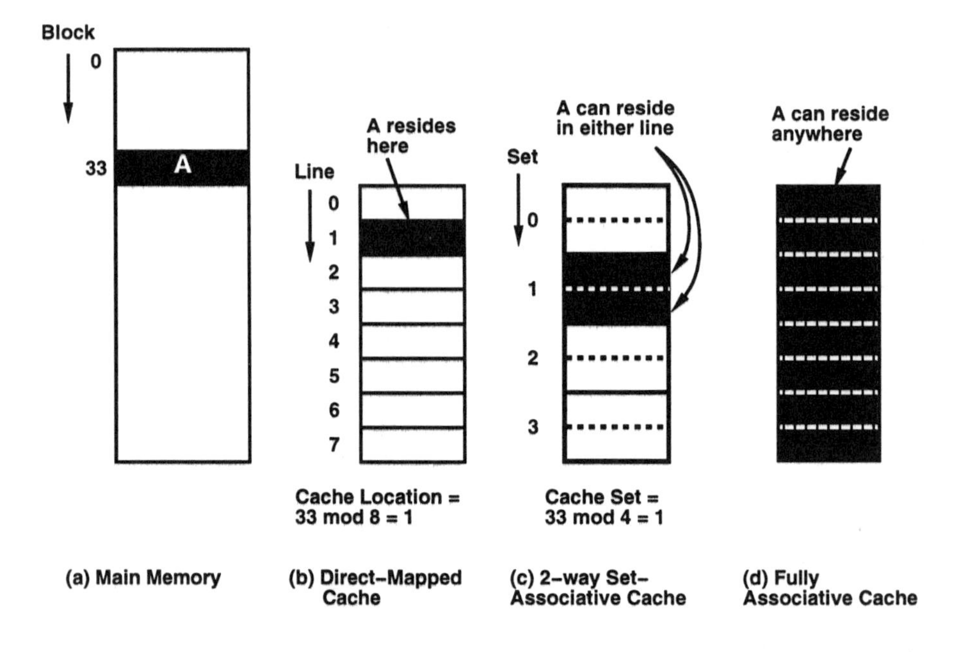

Figure 2.2. Mapping memory elements into the cache

Suppose we access two elements A and B in that order from the main memory. As shown in Figure 2.3(a), the two elements, which are located in memory blocks 33 and 41 respectively, map to the same address in the direct-mapped cache (Figure 2.3(b)), leading to a cache conflict, with B replacing A in the cache. Thus, if A were needed soon after, this leads to a cache miss. To alleviate problems of this nature, *set-associative caches* were devised. Here a memory block can be placed in any of a set of **a** cache lines, where **a** is the associativity. Thus, in the 2-way set-associative cache shown in Figure 2.2(c), A could reside in either of the two locations of the same set

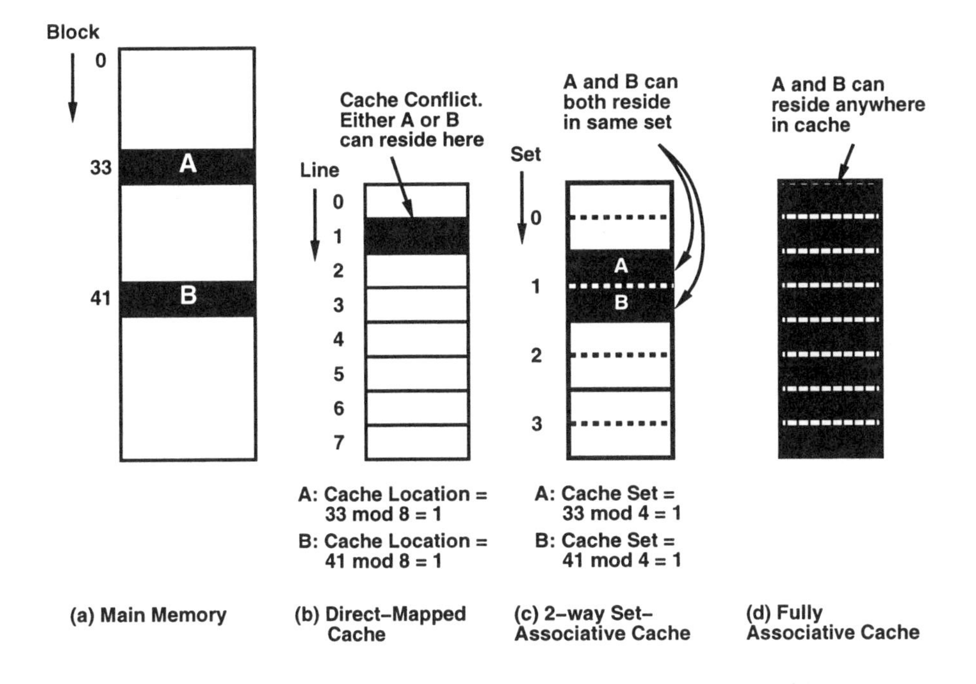

Figure 2.3. Cache Conflicts

shown, and in the generalized *fully associative cache* (Figure 2.2(d)), A could reside anywhere in the cache. Thus, as shown in Figure 2.3(c), if two elements A and B of memory are repeatedly accessed, they can both reside in the same cache set without incurring cache conflicts. In the fully associative cache (Figure 2.3(d)), cache conflicts are also avoided because A and B can reside anywhere in the cache.

While associative caches incur a lower miss ratio than direct-mapped ones by eliminating some cache conflicts, they are also more complex, and incur longer access times, because, now, **a** searches need to be performed to determine if a data element exists in the cache, as opposed to a single search in the case of direct-mapped caches. Conflict misses can be avoided by using a fully associative cache, but due to access time constraints, most cache memories employ a limited-associativity architecture. In fact, studies show that the simple direct-mapped cache architecture is indeed the best choice for most applications [Hil88]. For example, microprocessors including the HP 735, DEC AXP 3000, SGI Indigo2 and SUN Sparc5 are all equipped with direct-mapped data caches.

Further, there is additional hardware complexity in associative caches to implement a *replacement policy* to decide which cache line to evict from a cache set when a new cache line is fetched. In Figure 2.3, if another block C at memory address 49 (which also maps to set 1) were accessed, the replacement policy would help decide which of A and B is replaced by C. A common replacement policy is *Least Recently Used* (LRU) [HP94], in which the cache line that has not been accessed for the longest duration is replaced, in keeping with the principle of locality.

Other architectural techniques for improving hit ratio and reducing miss penalty are: *column-associative caches, victim caches*, and *hardware prefetching*.

2.2.1.3 Column-Associative Caches. A column-associative cache [AP93] has the same access time as direct-mapped caches on cache hits, but on misses, the cache also looks in a second cache line to check if the data resides there. If it does, then the data is returned, with the delay of an additional cycle, resulting in a *pseudo-hit*. Thus, memory accesses incur three types of delays, resulting from cache hits, pseudo-hits, and cache misses. The design of column-associative caches attempts to transform some possible cache misses into pseudo-hits.

2.2.1.4 Victim Caches. Another way to improve the hit ratio of a direct-mapped cache while retaining the same access time is to augment it with a small fully associative buffer called the *victim cache* [Jou90]. The victim cache stores recently evicted data from the main direct-mapped cache, so that, on a cache miss, the victim cache is first looked up before fetching data from main memory.

2.2.1.5 Hardware Prefetching. Hardware Prefetching refers to the idea of fetching data and instructions from memory even before the processor requests it [Jou90, HP94]. One common prefetching technique is to access the memory block requested, as well as the next contiguous block, in anticipation of spatial locality. The fetching of the next block is transparent to the processor, so that if the data has arrived by the time the processor requests it, a cache hit results.

2.2.2 Software Techniques

The earliest memory-related efforts in the compiler domain are those on the *register allocation* issue. Since the number of program variables is usually much larger than the number of physical registers, many variables would need to be stored in and accessed from main memory. Register allocation refers to the problem of allocating variables into a set of registers so as to minimize the number of accesses to main memory [ASU93]. Traditionally, register allocation is performed using a graph colouring technique [CAC+81]. Each program variable is represented by a node in a graph, with an edge between two nodes existing if the two corresponding variables can be allocated into the same register, i.e., the variables have non-overlapping *life-times* [ASU93]. The register allocation problem now becomes a graph colouring problem, with the number of colours equal to the number of available registers. If a legal colouring is not found, then some variables are *spilled* into main memory, and the process is repeated. In [KP87], the authors present REAL, a polynomial-time algorithm for optimal register allocation on a given basic block of code (i.e., code with no conditional branches). In the presence of conditional branches, optimal register allocation is an NP-complete problem. Register allocation in the presence of multiple register files was addressed in [CDN97]. A review of contemporary approaches to register allocation can be found in [Muc97].

Several compiler techniques exist for transforming code to exploit the underlying cache structure. One important principle of compiler optimization for improving cache performance is to access consecutive memory locations wherever possible to maximize cache hits when a block of memory elements is fetched into a cache line on a single cache miss (i.e., exploit spatial locality). Some important compiler optimizations that help improve cache performance are: *Loop Interchange, Loop Fusion, Loop Blocking, Loop Unrolling*, and *Software Prefetching*.

2.2.2.1 Loop Interchange. A simple example of cache-directed compiler transformation is Loop Interchange [BGS94], which is illustrated with the C-code in Figure 2.4. Assuming row-major data storage for the two dimensional array a, the code in Figure 2.4(a) exhibits poor spatial locality because the inner loop accesses elements along one column of the array, and consecutive column elements do not occur sequentially in memory. The loop interchange transformation of Figure 2.4(b) maintains the same behavior, while ensuring that elements along the same row are accessed in the inner loop, thereby resulting in improved cache performance arising out of better spatial locality.

2.2.2.2 Loop Fusion. The Loop Fusion transformation [BGS94] combines multiple loops into a single loop. This transformation improves cache performance by preventing multiple occurrences of cache misses to the same elements of the same array. Consider the two loops shown in Figure 2.5(a), with the assumption that the arrays a, b, and c are larger than the cache. Since array a occurs in both loops, the loading of a incurs cache misses in both loops. The loop fusion transformation shown in Figure 2.5(b) ensures that a is loaded only once into the cache. Further, $a[i]$ could be register-allocated, thereby eliminating the second memory access to array a in the loop.

2.2.2.3 Loop Unrolling. The Loop Unrolling transformation [BGS94] helps improve performance by reducing the number of accesses to the cache. Loop unrolling is illustrated in Figure 2.6, where the code in Figure 2.6(a) is unrolled once, resulting in

int $a[N][N]$;
⋮
for $i = 0$ **to** $N - 1$
  for $j = 0$ **to** $N - 1$
    $a[j][i] = a[j][i] + 1;$

(a) Original loop

int $a[N][N]$;
⋮
for $j = 0$ **to** $N - 1$
  for $i = 0$ **to** $N - 1$
    $a[j][i] = a[j][i] + 1;$

(b) Transformed Loop

Figure 2.4. Loop Interchange Transformation

for $i = 0$ **to** 1000 **step** 1
  $b[i] = a[i] + 1;$
for $i = 0$ **to** 1000 **step** 1
  $c[i] = a[i] * 3;$

(a) Original loop

for $i = 0$ **to** 1000 **step** 1
  $b[i] = a[i] + 1;$
  $c[i] = a[i] * 3;$

(b) Transformed Loop

Figure 2.5. Loop Fusion Transformation

the code of Figure 2.6(b). Since $a[i+1]$ is accessed twice in the unrolled loop body, it can be register-allocated, thereby saving an access to the cache. Unrolling also reduces the control overhead for the loop, but incurs the penalty of increased code size that, for large loop bodies, could adversely affect instruction cache performance.

2.2.2.4 Loop Blocking (or Tiling). Loop blocking (or tiling) is a compiler transformation that divides the iteration space into *blocks* or *tiles* to ensure that the data

for $i = 0$ **to** 99 **step** 1
$b[i] = a[i] + a[i + 1];$

(a) Original loop

for $i = 0$ **to** 49 **step** 2
$b[i] = a[i] + a[i + 1];$
$b[i + 1] = a[i + 1] + a[i + 2];$

(b) Transformed Loop

Figure 2.6. Loop Unrolling Transformation

for $i = 0$ **to** $N - 1$
for $j = 0$ **to** $N - 1$
for $k = 0$ **to** $N - 1$
$C[i][j] = C[i][j] + A[i][k] * B[k][j]$

(a) Original loop

for $j' = 0$ **to** $N - 1$ step TSIZE
for $k' = 0$ **to** $N - 1$ step TSIZE
/* for one TSIZE × TSIZE tile */
for $i = 0$ **to** $N - 1$
for $j = j'$ **to** $\min(j' + \text{TSIZE} - 1, N - 1)$
for $k = k'$ **to** $\min(k' + \text{TSIZE} - 1, N - 1)$
$C[i][j] = C[i][j] + A[i][k] * B[k][j]$

(b) Transformed Loop

Figure 2.7. Loop Blocking Transformation

set involved in a computation fits into the data cache [LRW91]. In Figure 2.7, the matrix multiplication code of Figure 2.7(a) is transformed into the tiled version of Figure 2.7(b), with the iteration space being divided into TSIZE × TSIZE tiles. The motivation for the transformation is that the arrays A, B, and C might be too large to all fit into the cache leading to too many capacity misses, hence, the tile sizes can be chosen so that the data accessed during computation involving a single tile can fit into the cache. We examine some strategies for intelligent selection of tile sizes in Chapter 4.

for $i = 0$ **to** $N - 1$
  for $j = 0$ **to** $N - 1$
    for $k = 0$ **to** $N - 1$
      LOAD $A[i][j]$, r_1
      LOAD $B[i][k]$, r_2
      LOAD $C[k][j]$, r_3
      $r_1 = r_1 + r_2 * r_3$
      STORE r_1, $A[i][j]$

(a) Original loop

for $i = 0$ **to** $N - 1$
  for $j = 0$ **to** $N - 1$
    for $k = 0$ **to** $N - 1$
      LOAD $A[i][j]$, r_1
      LOAD $B[i][k]$, r_2
      PREFETCH $B[i][k + 1]$
      LOAD $C[k][j]$, r_3
      PREFETCH $C[k + 1][j]$
      $r_1 = r_1 + r_2 * r_3$
      STORE r_1, $A[i][j]$

(b) Transformed Loop

Figure 2.8. Software Prefetching

2.2.2.5 Software Prefetching. Software Prefetching [CKP91] is a compiler technique for loading the data cache with data in advance of its actual use, which requires a *non-blocking cache load instruction* in the processor. The compiler attempts to predict in advance which data will be accessed in the future, and inserts the cache load instruction for the data at appropriate places in the code. This incurs a penalty of a single instruction, but could save many cycles by way of reduced cache misses.

Figure 2.8(a) shows an implementation of the matrix multiplication algorithm, with the load and store instructions explicitly shown. This code is modified with prefetch instructions as shown in Figure 2.8(b), using the following simple heuristic: if there is an array reference which is a function of inner loop index R in a loop body, then prefetch the same reference with $(R + 1)$ replacing R (or, in general, $R + s$, where s is the loop stride). Thus, in Figure 2.8(b), the prefetch of $B[i][k + 1]$ follows the access of $B[i][k]$ and the prefetch of $C[k + 1][j]$ follows the access of $C[k][j]$. (Note that k is the loop index in the inner loop). The prefetch instructions essentially fetch the data that is predicted to be accessed in the next loop iteration. Important issues involved in software prefetching include the ideal placement of the prefetch instruction [KL91, CMCH91], the decision of whether or not to prefetch [MLG92], and the possible use of a separate *prefetch buffer* (instead of the cache) to place prefetched data [KL91, CMCH91].

2.3 MEMORY ISSUES IN EMBEDDED SYSTEMS

Memory issues play an important role in the design of application-specific systems because many applications that access large amounts of data have to incorporate memory storage and memory accesses into the design. Studies have shown that in many Digital Signal Processing (DSP) applications, the area occupied and power consumed by the memory subsystem is up to ten times that of the datapath [BCM95], making memory a critical component of the design. One important difference between the role of memory in embedded systems and general-purpose processors is that, since there is only a single application in an embedded system, the memory structure is configurable, and can be tuned to suit the requirements of the given application. In Chapter 5, we demonstrate how the traditional memory hierarchy structure can be modified in the context of embedded systems resulting in improved performance. Further, while general-purpose systems are typically concerned only with the performance-related aspects of the memory system, embedded systems have a more diverse objective function, composed of:

- area,
- performance, and
- power consumption.

Memory area is also an important criterion in embedded systems because it is a significant contributor to the total chip area, which, in turn, determines the cost of the chip. Thus, several research efforts such as *memory allocation, memory packing, code size reduction* and *code compression* have targeted the minimization of memory area cost function.

Memory allocation is the problem of mapping behavioral variables into physical memory. Register allocation in compilers (Section 2.2.2) falls in this category. The scope of the register allocation problem can be broadened in the context of embedded systems. Recently, exact and heuristic techniques utilizing loop unrolling were proposed to perform optimal register allocation for variables in loops [KNDK96]. The exact algorithm is characterized by exponential time, but the heuristic algorithm also results in good performance with acceptable cost. Such algorithms, in spite of being time-consuming, deserve consideration in embedded systems because of the availability of longer compilation times.

While initial memory allocation efforts focussed on individual registers and register files, later efforts addressed the storage of variables in multiport memory. The first work in grouping variables to a multiport memory module while minimizing interconnection-related costs such as busses, multiplexers, and tri-state buffers, was presented in [BMB+88]. This approach was generalized to incorporate multiple memory modules in [AC91]. In the above approaches, the grouping of variables into memories preceded the determination of interconnections between functional units and memories. However, since the two stages are phase-coupled, the complexity of the interconnections cannot be independently optimized. In [KL93], the authors took the approach of minimizing interconnection cost first, followed by grouping of variables into multiport memories.

The PHIDEO synthesis system, which is targeted at high-speed pipelined DSP applications, uses a slightly different memory model in that registers (foreground memory) are not differentiated from memory modules (background memory). Memory modules occur as buffers between processing units. The input for its memory allocation tool MEDEA is in the form of hierarchical (multi-dimensional) data streams. The tool allocates data streams to different memory modules, while attempting to minimize interconnection costs between memories and functional units. In [BCM94], a *memory allocation* and assignment technique for video image processing systems is presented. A data-flow driven analysis of a non-procedural specification in the SILAGE language is employed to determine the type, word-length, size, and port requirements for memories. This memory allocation technique incorporates *in-place optimization* (the possibility of reusing the same memory space for different sections of the same multidimensional signal) using a polyhedral data-flow analysis.

When memories in embedded systems are composed of smaller units from a library of components, there arises the problem of *memory packing* – how to organize smaller physical memory modules to realize the required logical memories for an application, while satisfying performance constraints. In [KR94], a branch-and-bound algorithm that considers a technology-dependent delay model is presented. The delay model and packing algorithm are oriented towards a Field Programmable Gate Array (FPGA) architecture. In [JD97], the authors present an Integer Linear Programming (ILP) based technique for solving the packing problem. A solution to the *memory selection* problem, where a suitable organization of memory modules is selected from a library that satisfies pipeline latency and throughput constraints, is presented in [BG95].

Size reduction of compiler-generated code has been a topic of recent interest in the embedded system research community. The motivation here is that, if the code is to be stored in on-chip ROM, then the ROM size contributes to the total chip area, thereby affecting cost. Since many embedded systems employ DSP cores, researchers have focussed their attention on code size reduction for processors such as the TMS320 DSP processor family from Texas Instruments. Traditional compilers for general-purpose processors usually generate sub-optimal code for DSPs because of the presence of irregular datapaths, specialized registers, and limited connectivity in DSPs. One related problem is that of memory address assignment. Many DSPs offer the auto-increment and auto-decrement feature for sequential access to memory data. If variables are suitably placed by the compiler in memory to effectively utilize auto-increment and auto-decrement modes, then code size can be reduced by eliminating unnecessary increment and decrement instructions for updating the memory address register. Techniques for variable address assignment to exploit these modes are discussed in [LDK+95, LM96, Geb97, LPJ96]. Code compression is another strategy that helps reduce memory area in embedded systems. Compressed code is stored in the main memory, and the code is decompressed one cache line at a time when it is fetched into the cache [LW98, BS97].

An important task in the memory design of embedded systems is the *estimation* of the area and performance of candidate memory configurations to enable fast comparisons. Strategies for estimating the total on-chip area required to store the data for a given application, accounting for in-place optimization, have been studied in [BCM95, VSR94]. The estimation problem is also relevant from the performance point of view. In the CINDERELLA system [LMW95], the worst case execution time (WCET) for a program is computed using an instruction cache model and an ILP formulation. The program is divided into blocks equal to the cache line size, and the number of hits and misses in the direct-mapped cache is estimated from an instruction cache conflict graph. In the POLIS system [SSV96], the effects of cache are incorporated into the performance estimation step of hardware/software codesign by explicit simulation.

Cache memory issues have been the subject of recent study in the area of embedded systems. Early work on the issue of code placement in main memory to maximize instruction cache hit ratio [McF89] was extended recently in [TY97]. Although this technique succeeds in increasing the hit ratio, thereby improving performance, this is at the expense of area, since the code size increases in the effort to obtain favourable placement of instruction blocks. In [TY96b] the authors present a variation of the above technique that incorporates a given code size constraint while attempting to maximize the instruction cache hit ratio. A model for partitioning the instruction cache among several processes was presented in [LW97].

With power consumption rapidly becoming an important objective function in embedded systems, some recent efforts have focussed on the possibility of power reduction during memory accesses. Works such as [MLC97, SB95, BdMM+98] have concentrated on encoding the memory address and data busses in order to reduce I/O transitions, thereby reducing power consumption resulting from switching of high-capacitance wires. In [TIIY98], the authors have presented an instruction scheduling technique for power reduction in instruction caches, which seeks to minimize signal transitions between consecutive instructions fetched from memory. In [LH98], the authors present an algorithm for trading off energy dissipated by the instruction execution unit against that dissipated at the cache/main memory interface. In [KCM98], the authors outline a methodology for program optimizations targeting power reduction in the cache that incorporates the in-place optimization and memory allocation of [WCF+94, GCM98]. In [WCM95], the authors have addressed the problem of mapping set data structures used in networking applications into memory so as to reduce power consumption when the data is accessed.

The various different memory-related issues in embedded systems, coupled with the numerous ways in which memory can be configured to fulfill the requirements of an application, makes it necessary to address the issue of *memory exploration* in system design in order to determine an efficient memory configuration that satisfies given constraints and optimizes area, performance, and power consumption metrics. A system level memory exploration technique for ATM applications is presented in [SWCdJ97], where the dominance of the memory subsystem in both area and power consumption leads them to perform the memory optimizations as a first step in system design. An important subtask in the exploration is the minimization of the required memory bandwidth (i.e., maximum number of simultaneous accesses in the same cycle) in an implementation, which ultimately results in a reduction in the number of memory modules and the number ports in the generated memory architecture. In [KLPMS97], the authors present a simulation-based technique for selecting a processor and required instruction and data caches.

2.4 IN THIS BOOK

The subject of memory-related optimization and exploration forms the core of this book. We study memory-related optimizations both from the viewpoint of performance as well as power consumption. A major portion of the book is devoted towards the incorporation of traditional memory hierarchy concepts presented in Section 2.2 into the embedded system synthesis field summarized in Section 2.3. As we shall see, the considerable knowledge of the application-specific environment allows many architectural innovations and compiler optimizations that would not be valid in a general-purpose processor scenario. Further, the availability of longer compilation times affords a deeper analysis during the memory exploration of embedded systems.

3 OFF-CHIP MEMORY ACCESS OPTIMIZATIONS

Due to the widening gap between processor and memory speeds, the memory interface has become a serious performance bottleneck because off-chip memory access is a relatively slow operation, incurring 10-20 times the delay of a typical datapath operation [PH94]. The consideration of memory issues is critical because, in many applications such as signal processing, memory storage dominates chip area and power – typical on-chip memory area is reported to be almost 10 times the typical area of a complex datapath [BCM95]. Thus, it becomes imperative to consider the effects of memory accesses during synthesis in order to realize efficient system designs. In this chapter, we address performance and power optimization issues in the interface between the synthesized hardware and off-chip memory blocks of the embedded system architecture, which are highlighted in Figure 3.1. In Section 3.1 we describe presynthesis transformations and optimizations that allow high-level synthesis (HLS) to effectively utilize contemporary off-chip memory modules such as DRAMs. In Section 3.2 we describe techniques to minimize address bus transition activity with the goal of reducing power dissipation during off-chip memory accesses.

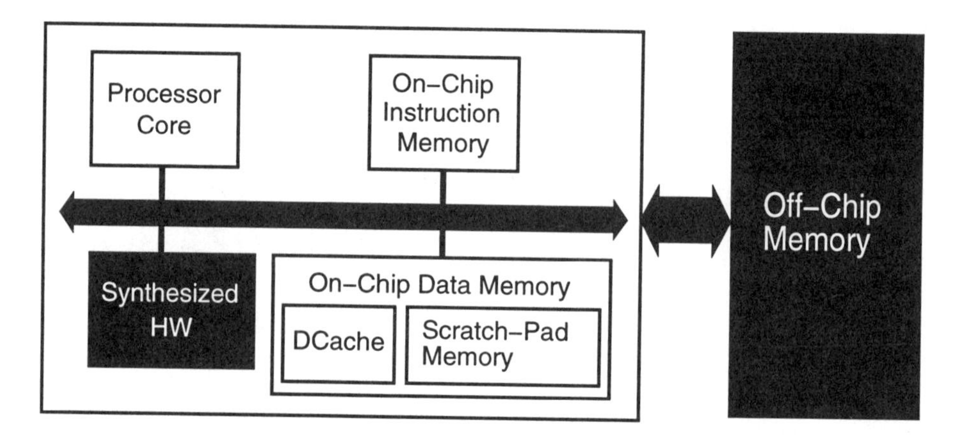

Figure 3.1. Shaded blocks represent the synthesized hardware/off-chip memory interface

3.1 INCORPORATING OFF-CHIP MEMORY ACCESSES INTO SYNTHESIS

Arrays in behavioral specifications are typically assigned to memories during synthesis. If these arrays are small enough, they may be mapped into on-chip SRAM. However, many applications involve large arrays, which need to be stored in off-chip memories, such as DRAM. Consequently, it is essential to employ a reasonably accurate model for memory operations during synthesis. Modern memories have efficient access modes (such as *page mode* and *read-modify-write*), that are known to improve the access bandwidth [Fly95]. Since the newer generation of memories (Extended Data Out DRAMs, Synchronous DRAMs, etc.) all incorporate these access modes, it is important to incorporate such realistic memory library interface protocols into high-level synthesis, so that scheduling, allocation and binding algorithms can exploit these features to generate faster and more efficient designs.

In order to incorporate off-chip memory accesses into current synthesis tools and algorithms, we would have to either model all memory accesses as multicycled operations and use a standard DRAM controller interface to the off-chip memory (as is the case with typical Processor-Memory interfaces); or use a more refined multi-stage *behavioral template* model [LKMM95]. A behavioral template represents a complex operation in a Control Data Flow Graph (CDFG) by enclosing several individual CDFG nodes, and fixing their relative schedule. Templates allow a better representation of memory access operations than single multicycled CDFG nodes, and lead to better schedules. However, since different memory operations are treated independently, typical features of realistic memory modules (e.g., page mode for DRAM) cannot be exploited, because the scheduler has to assume the worst-case delay for every memory operation, resulting in relatively longer schedules. Further, system designers will not use synthesis tools that do not effectively exploit various memory access modes and protocols. In order to utilize these efficient memory modes, the synthesis algorithms have to incorporate them *during* or *before* scheduling.

In this section, we examine the following issues related to the incorporation of off-chip memory accesses in high-level synthesis:

1. We describe models for the well-known operation modes of off-chip memories (DRAMs), which can be effectively incorporated into HLS tools.

2. We describe algorithms for inferring the applicability of the memory access modes to memory references in the input behavior.

3. We outline techniques for transforming the Control-Data Flow Graph (CDFG) for the input behavior to incorporate the memory access modes, so as to obtain an efficient schedule.

The technique we describe serves as presynthesis transformations and optimizations for high-level synthesis, so that a typical scheduling algorithm, such as list scheduling, can be efficiently applied to the transformed CDFG.

3.1.1 Memory Access Optimizations in DRAMs

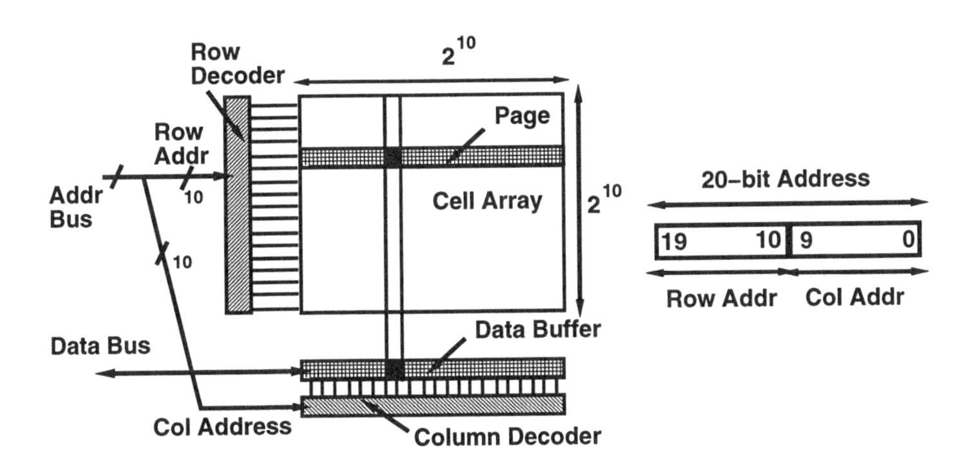

Figure 3.2. Typical DRAM Organization

Figure 3.2 shows a simplified view of the typical organization of a DRAM with 2^{20} words (= 1 M words), with the core storage array consisting of a $2^{10} \times 2^{10}$ square. A

20-bit logical address for accessing a data word in the DRAM is split into a 10-bit *Row Address* – consisting of the most significant 10-bits (bits 19 . . . 10); and a 10-bit *Column Address* – consisting of the least significant 10 bits (bits 9 . . . 0). The *Row Decoder* selects one of 2^{10} rows (or DRAM *pages*[1]) using the row address, and the *Column Decoder* uses the column address to select the addressed word from the selected page. The following are some well-known features of the memory organization that result in savings of pin count and efficiency of accesses.

- Since the row and column addresses are used at different points of time, the same address pins are used to time-multiplex between row and column addresses. The address bus input to the DRAM is only 10 bits wide. This multiplexing reduces the number of address bits, resulting in reduced off-chip pin count.
- The row-decoding step physically copies an entire page into a data buffer (Figure 3.2), anticipating spatial locality, i.e., expecting future references to be from the same page. If the next memory access is to a word in the same page, the row-decoding phase can be omitted, leading to a significant performance gain. This mode of operation is called *page mode*.

Modern DRAMs commonly utilize the following six memory access modes:

Read Mode – single word read, involving both row-decode and column-decode.

Write Mode – single word write, involving both row-decode and column-decode.

Read-Modify-Write (R-M-W) Mode – single word update, involving read from an address, followed by write *to the same address*. This mode involves one row-decode and column-decode stages each, and is faster than two separate Read and Write accesses.

Page Mode Read – successive reads to multiple words in the same page.

Page Mode Write – successive writes to multiple words in the same page.

Page Mode Read-Modify-Write – successive R-M-W updates to multiple words in the same page.

In order to motivate the need for explicitly modeling and exploiting these modes, we demonstrate the effect on design performance in HLS between scheduling with normal read operation, versus scheduling that exploits the page mode read of a DRAM. The sample library memory module used is the IBM11T1645LP Extended Data Out (EDO)

[1] For simplicity of explanation, we assume that one page = one row.

DRAM [IBM96], and the input behavior is the *FindAverage* routine in Figure 3.4(a), where the scalar variable *av* is mapped to an on-chip register, and the array $b[0\ldots3]$ is stored in off-chip memory.

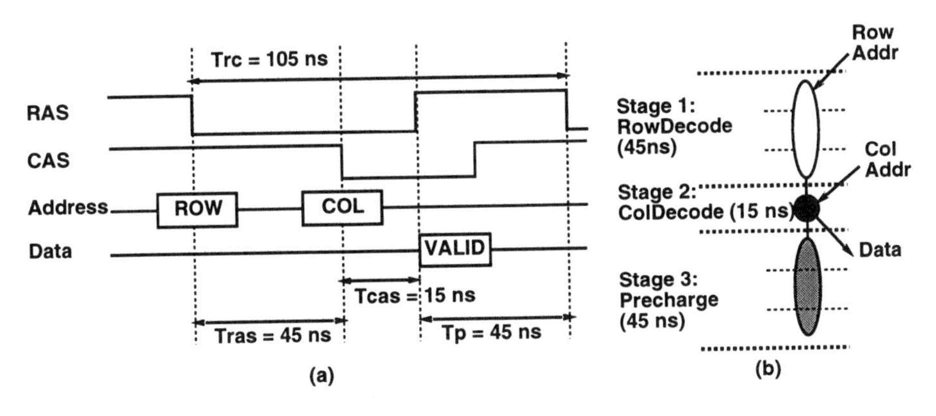

Figure 3.3. (a) Timing diagram for Memory Read cycle (b) Model for Memory Read operation

Figure 3.3(a) shows a simplified timing diagram of the *read cycle* of the $1M \times 64$-bit EDO DRAM. We use the timing characteristics of this chip for illustration throughout this section. The Memory Read cycle is initiated by the falling edge of the RAS (Row Address Strobe) signal, at which time the row address is latched from the address bus. The column address is latched later at the falling edge of CAS (Column Address Strobe) signal. Following this, the data is available on the data bus after a delay. Finally, the RAS signal is held high for long enough to allow for *bit-line precharge*, which is necessary before the next cycle can be initiated.

Table 3.1. Timing values for Read cycle

Name	Description	Value
T_{rac}	Access time from RAS edge	60 ns
T_{cac}	Access time from CAS edge	15 ns
T_{rc}	Read cycle time	105 ns
T_{ras}	Row Decode time	$60 - 15 = 45$ ns
T_{cas}	Column Decode time	15 ns
T_p	Precharge (Setup) time	$105 - 60 = 45$ ns

Table 3.1 summarizes the important timing characteristics for the read cycle. T_{rac} gives the minimum delay between the falling edge of RAS and the availability of data, and T_{cac} gives the delay between the CAS edge and the data availability. Logically,

T_{cas} ($= T_{cac}$) is the time for column address decoding, and T_{ras} ($= T_{rac} - T_{cac}$) is the row address decoding time, which also includes the time for transfer of the memory page into the data buffer. The total read cycle time, T_{rc}, is significantly larger than T_{rac}, because the precharging of the memory bit lines (setting them up for the next memory operation) is a time-consuming operation. This setup time (T_p) can be calculated as $T_p = T_{rc} - T_{rac}$, since that is the extra time spent after data is available.

From the above timing characteristics, we can derive a CDFG node cluster for the memory read operation consisting of 3 stages (Figure 3.3(b)):

1. Row Decode

2. Column Decode

3. Precharge

The row and column addresses are available at the first and second stages respectively, and the output data is available at the beginning of the third stage. Techniques for formally deriving the node clusters from interface timing diagrams have been studied in the interface synthesis works such as [COB95, Gup95], and can be applied in this context.

Assuming a clock cycle of 15 ns, and a 1-cycle delay for the addition and shift operations, we derive the schedule shown in Figure 3.4(b) for the code in Figure 3.4(a), using the memory read model in Figure 3.3(b). Since the four accesses to array *b* are treated as four independent memory reads, each of these incurs the entire read cycle delay of $T_{rc} = 105$ ns, i.e., 7 cycles, requiring a total of $7 \times 4 = 28$ cycles.

However, DRAM features such as page mode read can be efficiently exploited to generate a tighter schedule for behaviors such as the *FindAverage* example, which access data in the same page in succession. Figure 3.4(c) shows the timing diagram for the page mode read cycle, and Figure 3.4(d) shows the schedule for the *FindAverage* routine using the page mode read feature. Note that the page mode does not incur the long row decode and precharge times between successive accesses, thereby eliminating a significant amount of delay from the schedule. In this case, the column decode time is followed by a *minimum pulse width* duration for the CAS signal, which is also 15 ns in our example. Thus, the effective cycle time between successive memory accesses has been greatly reduced, resulting in an overall reduction of 50% in the total schedule length.

The key feature in the dramatic reduction of the schedule length in the example above is the recognition that the input behavior is characterized by memory access patterns that are amenable to the page mode feature, and the incorporation of this observation in the scheduling phase. In the following sections, we first describe synthesis models for the various access modes available in modern off-chip memories,

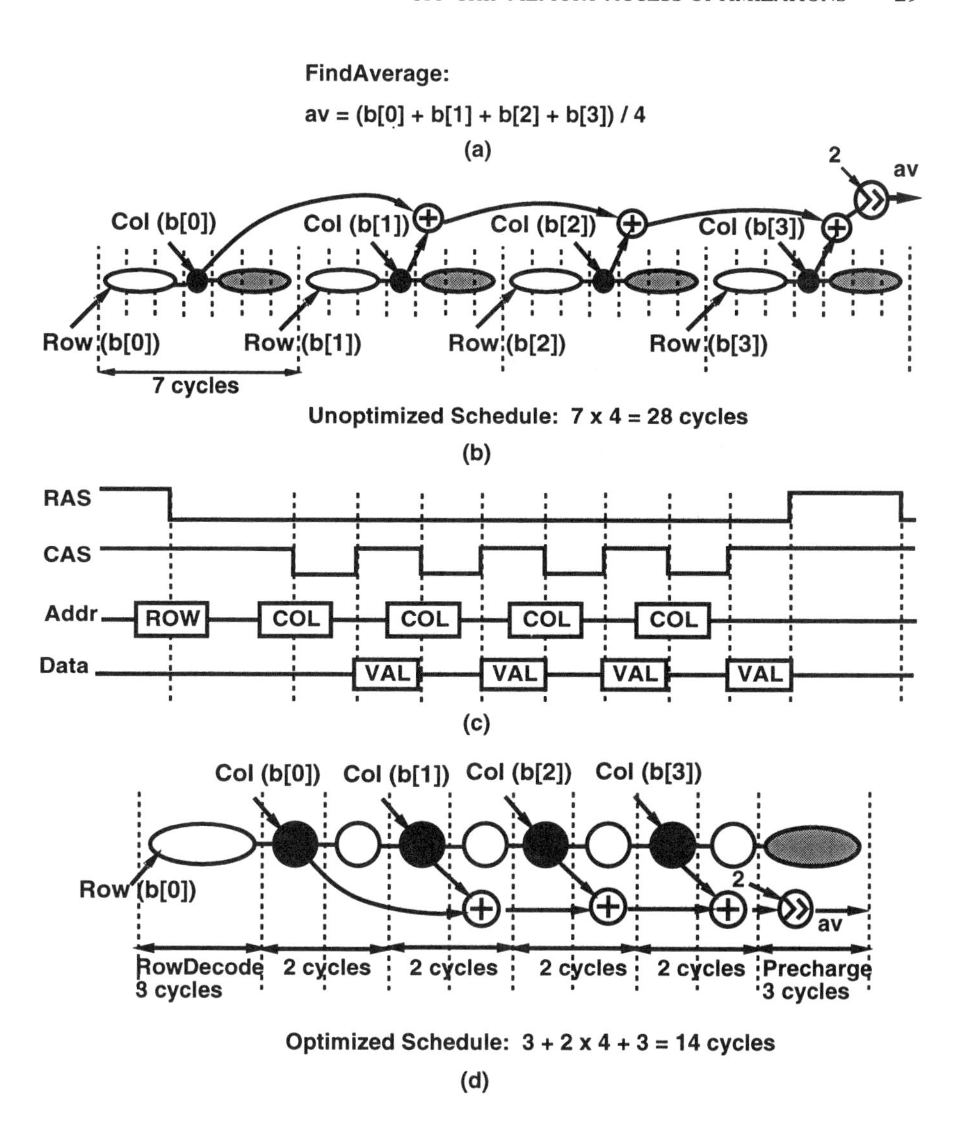

Figure 3.4. (a) Code for *FindAverage* (b) Treating the memory accesses as independent Reads (c) Timing diagram of *page mode read* cycle (d) Treating the memory accesses as one page mode read cycle

followed by a technique that incorporates the models into HLS by transforming the input behavior accordingly.

3.1.2 Representing Memory Accesses for HLS

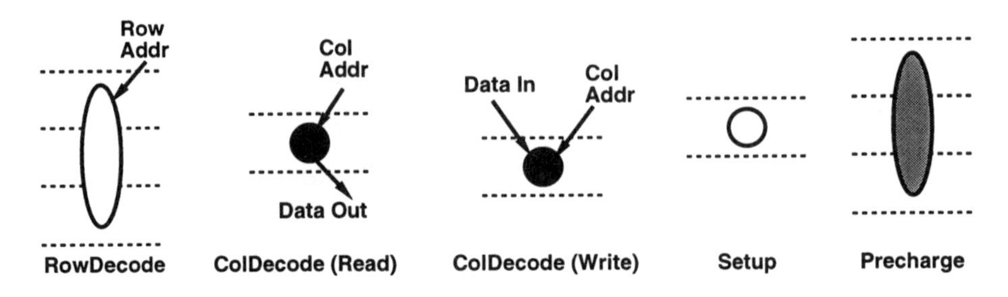

Figure 3.5. CDFG nodes used for memory operations

Instead of a single memory access node in the CDFG, we model memory operations corresponding to the different access modes using finer-grained memory access nodes (Figure 3.5).

- The **RowDecode** node is a multicycle operator that marks the beginning of any memory operation. The node has one input – the row address of the location, which should be ready before this node is scheduled.
- The **ColDecode (Read)** and **ColDecode (Write)** nodes form the second stage of the memory read and write operations. The **ColDecode (Read)** node has one input – the column address, and one output – the data read. The **ColDecode (Write)** node has two inputs – the column address, and the data to be written.
- The **Setup** node serves as a "delay" node in order to implement minimum delay constraints between successive stages of a memory operation.
- The **Precharge** node is a multicycle node that marks the last stage of a memory cycle. Physically, it signifies the restoration of the memory bit-lines to the initial state, so that the next operation can be initiated after the completion of this stage.

For each memory access mode, we build composite memory access nodes in the CDFG, based on the access protocol for that mode.

Read Mode: This mode was described in Section 3.1.1.

Write Mode: The memory write operation has similar timing characteristics to the read operation. The CDFG node cluster for the write cycle is shown in Figure 3.6(a). The ColDecode node used in the second stage is the **ColDecode (Write)** node identified in Figure 3.5.

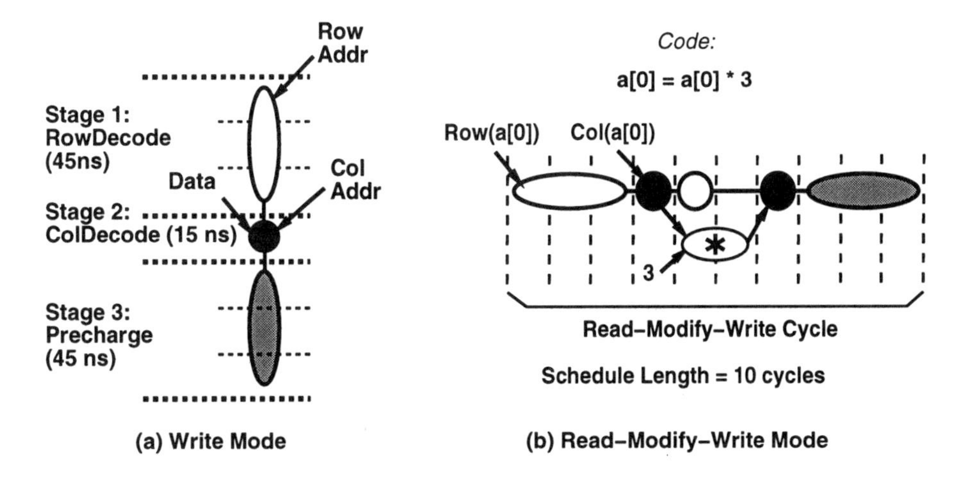

Figure 3.6. CDFG models of memory operations

Read-Modify-Write Mode: The Read-Modify-Write (R-M-W) mode is illustrated in Figure 3.6(b) with a simple behavioral statement: $a[0] = a[0] * 3$, which involves the reading and writing of memory address for $a[0]$. The schedule requires 10 cycles. Note that an extra control step is introduced between the *setup* and *write* stages, because the '$*$'-operation requires 2 cycles. If the operation were to be modeled as separate read and write cycles, we would require: $2 \times 7 = 14$ cycles.

Page Mode Read: This mode was described in Section 3.1.1.

Page Mode Write and **Page Mode R-M-W** modes are constructed similarly.

3.1.3 Incorporation of Memory Models in HLS

The memory operations described in Section 3.1.2 were designed for typical behavioral memory access patterns. For example, the page mode operations result in significant performance improvements in loops accessing arrays indexed by the loop variable. However, to utilize the memory features, we have to first analyze the behavioral specification to determine the applicability of the various efficient modes, and incorporate them during scheduling. To achieve this, we analyze the CDFG to identify behavioral patterns that can be optimized by the various efficient memory access modes (such as page mode), and transform it to incorporate the optimizations. The actions that need to be performed include:

Clustering of scalar variables into off-chip memory pages.

Reordering of memory accesses in the CDFG to exploit efficient memory operations (such as R-M-W).

Hoisting conditional memory operations prior to the evaluation of the condition, in order to shorten the schedule length.

Loop transformations in the CDFG to utilize page mode operations efficiently.

We describe the above actions in detail below.

3.1.3.1 Clustering of Scalars. Scalar variables are normally assigned to on-chip registers. However, if the number of such variables is large, it might be necessary to store these variables in memory. The scalar variables stored in off-chip memory have to be assigned memory addresses. A related optimization problem that arises in this address assignment is that, consecutive accesses to two different scalar variables can be implemented as a single page mode operation if both are located in the same memory page. Suppose the off-chip memory has a page size of P words. The following problem needs to be solved:

Group the scalar variables into clusters of size P (to reside in the same memory page), such that the number of consecutive accesses to the same memory page is maximized.

A similar problem has been addressed in the context of cache memory, where scalar variables are grouped into clusters of size L (where L is the size of a *cache line*), so as to minimize the number of cache misses (Chapter 4). When L variables are grouped in such a fashion, access to one of them causes all the L variables to be fetched into the cache, thereby reducing the number of cache misses. Since this problem has exactly the same formulation as our problem of clustering variables to fit into the same memory page, we use the same technique outlined in Chapter 4 for the clustering problem.

At the end of this step, each variable is assigned a memory address. We assume, for convenience, that the first elements of all arrays are aligned to a memory page boundary. Similarly, each row of a multi-dimensional array is padded so that all rows begin at a page boundary, unless the array is small enough to be accommodated within the same page.

3.1.3.2 Reordering of Memory Accesses. The correct ordering of memory accesses is critical for exploiting efficient memory access modes such as R-M-W. For example, in the code: "$a[i] = b[i] + a[i]$", the sequence of accesses: "*Read* $b[i] \rightarrow$ *Read* $a[i] \rightarrow$ *Write* $a[i]$" allows the utilization of the R-M-W mode for the address $a[i]$, while the sequence "*Read* $a[i] \rightarrow$ *Read* $b[i] \rightarrow$ *Write* $a[i]$" does not allow the mode, because of the intervening "*Read* $b[i]$" operation.

The possibility of utilizing R-M-W mode can be identified by starting from each "*Read(x)*" memory access (where x is a scalar variable or an array element), and proceeding down the DFG to find a path to the next *"Write(x)"* access, if it exists. If multiple paths exist, we combine them all together and still call the combination a "path". We limit this search to a single basic block (a section of the behavior that does not include any conditional statement).

However, there could be conflicting R-M-W paths that cannot be simultaneously satisfied. An example of such a memory access path is the example shown in Figure 3.7(a), for which the corresponding DFG is shown in Figure 3.7(b). Figure 3.7(b) also shows three possible R-M-W paths, for addresses $a[i]$, $b[i]$, and $c[i]$. Note that it is not possible to implement the memory accesses to all the three locations as R-M-W operations. For example, if we implement the $c[i]$ path as an R-M-W operation, then the $b[i]$ path can no longer be implemented as R-M-W, because *"Read* $b[i]$*"* has to occur before *"Write* $c[i]$*"*, thus forcing the $b[i]$ path to be split into separate *"Read* $b[i]$*"* and *"Write* $b[i]$*"* operations.

Thus, if two *"Read*(x) $\rightarrow$ *Write*(x)*"* paths intersect in the DFG, then both cannot be implemented as R-M-W operations. In the example DFG shown in Figure 3.7(b), if the $b[i]$ path is implemented as an R-M-W operation, it forces the $c[i]$ and $d[i]$ paths to be broken into individual read and write operations. The schedule resulting from this, shown in Figure 3.7(c), has a length 38 cycles. However, if the $c[i]$ and $d[i]$ paths are implemented as R-M-W (since $c[i]$ and $d[i]$ paths do not intersect), only the $b[i]$ path is broken, resulting in the schedule of Figure 3.7(d), which is only 32 cycles long.

In order to minimize the schedule length for a DFG, we need to solve the following optimization problem:

For a given DFG corresponding to a basic block in the behavior, with a set U of n different "Read(x) $\rightarrow$ Write(x)" paths, select a maximally non-intersecting subset $V \subseteq U$, i.e., find a subset V, such that all paths in V are non-intersecting, and $|V|$ is maximized.

The problem above can be shown to be NP-complete by building a graph G where each node corresponds to each path in U, and an edge (x, y) exists in G if the corresponding paths x and y in U are non-intersecting. The problem of determining the maximally non-intersecting set is equivalent to the problem of finding the maximal clique (fully connected subgraph) in the graph G, which is known to be NP-complete [GJ79]. Figure 3.7(e) shows the graph G derived from the DFG with paths $b[i]$, $c[i]$, and $d[i]$ of Figure 3.7(b). The maximal clique in the graph is indicated by the shaded region. Thus, an optimal reordering of the memory accesses for the DFG would be to implement the $c[i]$ and $d[i]$ paths as R-M-W operations, and break the $b[i]$ path into separate read and write cycles.

To obtain an approximate solution to the maximally non-intersecting subset problem, we build graph G as above, and apply a greedy heuristic, where we start with set $V = \phi$, and iteratively add node y to set V, such that y is the node in $U - V$, that has edges to all nodes in V, with highest incident number of edges in graph G.

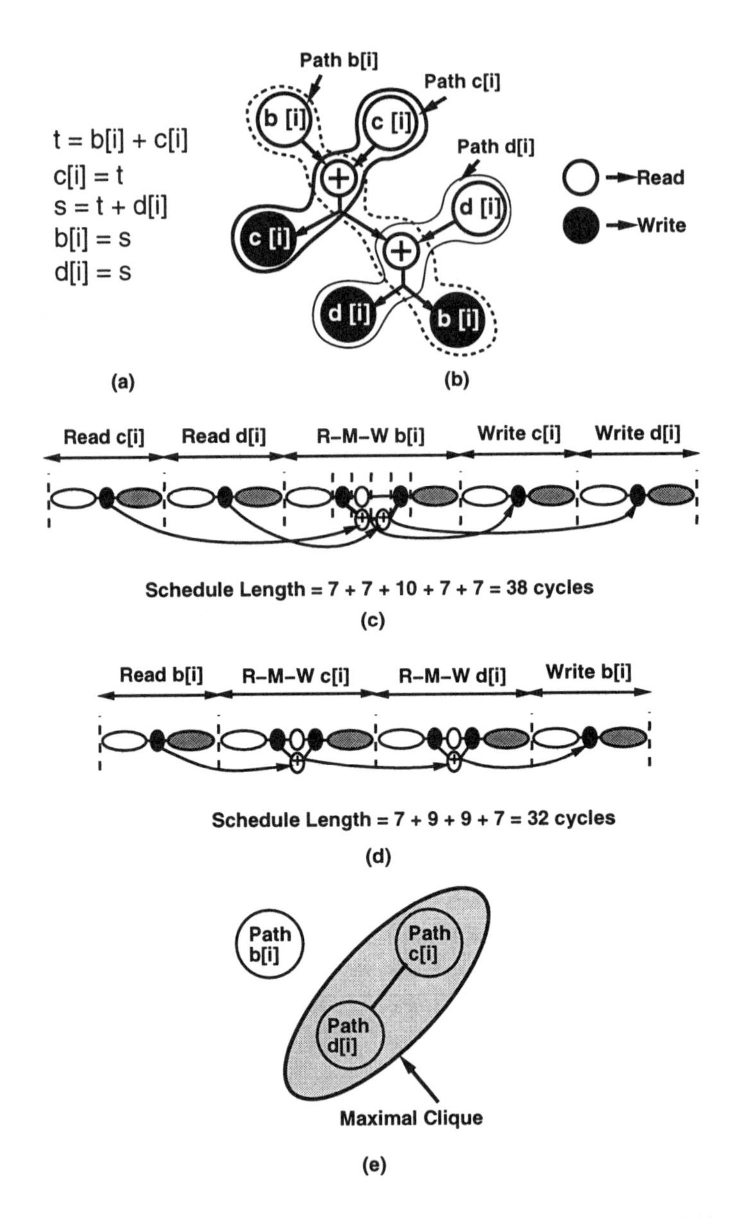

Figure 3.7. Ordering of memory operations determines possibility of exploiting R-M-W mode. (a) Sample code segment (b) DFG for code (c) Using R-M-W for $b[i]$ path (c) Using R-M-W for $c[i]$ and $d[i]$ paths (e) Graph G showing non-intersecting paths in DFG. Paths constituting the maximal clique are implemented as R-M-W operations

The heuristic has complexity $O(n^2)$, where n is the number of memory operations in the DFG. We mark all the paths in the selected set V to be implemented as R-M-W operations. However, it should be noted that, since the number of such paths are very small in typical behaviors, an exhaustive solution can also be used.

In order to execute the R-M-W cycle, it is necessary that all other memory read operations of the DFG, on which the R-M-W cycle depends, need to be completed first. To convey this information to the scheduler, we introduce *precedence edges* [GDLW92] in the CDFG to the *Read* node (i.e., first stage) of the R-M-W path T under consideration, from all other *Read* nodes that have a path leading to any node in path T. The *Read* and *Write* nodes in the DFG for a basic block, that do not form part of any identified R-M-W operation, can be further reordered to exploit page mode read and write modes. Since we have the address information at this stage, we divide the accesses in the basic block into groups of accesses from the same memory page.

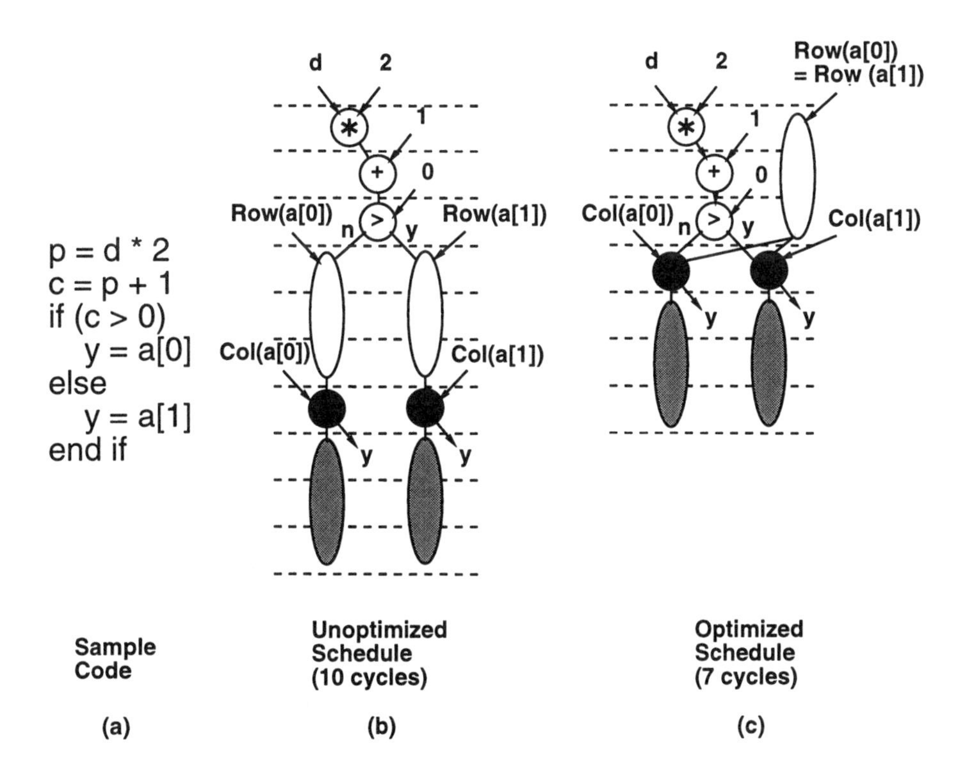

Figure 3.8. Hoisting Optimization with conditional memory Read in different branches

3.1.3.3 Hoisting. Due to the time-multiplexing of the memory address bus between row and column addresses, a scheduling optimization is possible when two addresses in the same memory page are accessed from different paths of a conditional branch. Consider the behavior shown in Figure 3.8(a). Assume that variables p, c, d, and y are stored in on-chip registers, whereas array a is stored in off-chip memory. Either $a[0]$ or $a[1]$ is fetched from memory, depending on the result of the conditional evaluation $(c > 0)$. A simple schedule, using the 3-stage read cycle model of Figure 3.3(b), and the assumption that $+, *$, and $>$ operations require 1 cycle, results in a schedule of length 10 cycles (Figure 3.8(b)). However, the knowledge that $a[0]$ and $a[1]$ reside in the same memory page (memory address assignments are statically computed) allows us to infer that both have the same row address, and hence, the read cycle could be initiated before the comparison operation. The schedule resulting from this optimization results in a length of only 7 cycles (Figure 3.8(c)). Output y is available after 7 cycles in the normal schedule, but after only 4 cycles in the optimized schedule. We call this optimization *hoisting* of the RowDecode stage. This is analogous to the speculative code hoisting optimization performed by compilers.

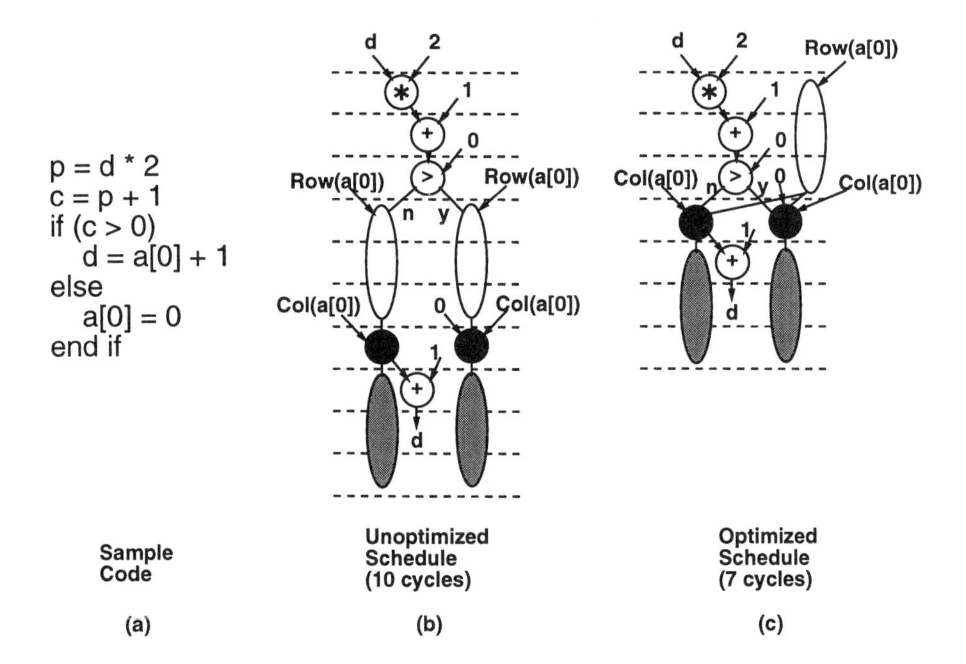

Figure 3.9. Hoisting Optimization with conditional memory Read and Write in different branches

Clearly, the *hoisting* optimization is also applicable to the write cycle when memory writes are in different conditional paths. A further scheduling optimization possible in this context is that, the decision about whether to initiate a read or write cycle can be deferred until the second stage of the cycle (since the WE signal to the memory, which distinguishes *Read* from *Write*, is needed only in the second stage). In the code shown in Figure 3.9(a), one path in the conditional ($c > 0$) leads to a memory write, whereas the other leads to a read. The normal schedule, which waits for the result of the comparison before choosing which cycle to initiate, results in a length of 10 cycles (Figure 3.9(b)). However, the deferred read vs. write feature can be used to initiate the cycle ahead of the time by issuing the row address (i.e., scheduling the RowDecode node) even before the condition is evaluated. The effect of this optimization is shown in Figure 3.9(c), where the result of the condition is used as the WE input signal to the second stage, where the appropriate operation is activated. This optimized schedule results in a length of only 7 cycles. This optimization is another form of *hoisting*.

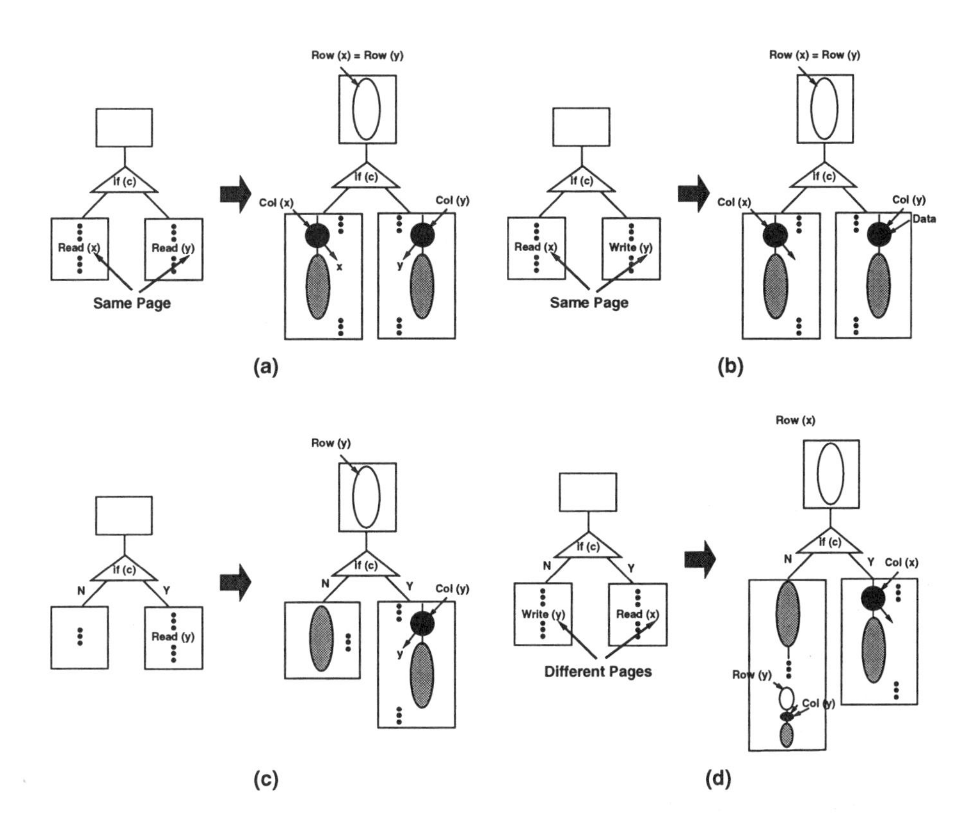

Figure 3.10. (a) & (b) Hoisting of RowDecode node when x and y are in the same page (c) Hoisting when only one branch has memory accesses (d) Hoisting when branches have accesses to elements x and y in different pages

In order to incorporate the above two hoisting transformations into the CDFG of the input behavior, we examine both branches of an *if* node, and select a pair of elements (x, y), such that x and y are accessed (read or written) in different branches, and the accesses to both are not data-dependent on any other memory accesses in the same basic block. If x or y is part of an R-M-W operation (identified in Section 3.1.3.2), we hoist the RowDecode stage *only if none of the nodes in the R-M-W path has any data dependence on another memory access*. This is done in order to prevent any R-M-W path from being broken due to the hoisting.

The CDFG transformations for the two types of hoisting, corresponding to the optimizations in Figures 3.8 and 3.9, are shown in Figure 3.10(a) and (b). In both cases, the memory accesses are replaced by the corresponding node clusters for read and write, with the RowDecode node being hoisted past the conditional.

Two other similar transformations are illustrated in Figures 3.10(c) and (d). Figure 3.10(c) shows a CDFG involving a conditional, in which there is a memory access (*Read (y)*) in one branch (labelled Y), and no memory access in the other (labelled (N)). In this case, the RowDecode node for *Read(y)* can be issued before the conditional is evaluated, resulting in a possibly faster schedule for the Y branch. However, since this might lead to an incorrect memory state if the N branch is taken, we insert a Precharge node into the N branch, so that the memory is now correctly reset for future operations. This is necessary because the bit-lines in the memory, which were discharged in the RowDecode (Section 3.1.1), now have to be precharged back before the next memory cycle can be initiated. In order to not penalize the N branch with additional delay, we perform this transformation only if the following equations hold:

$$t \geq P \tag{3.1}$$

$$c \geq RD \tag{3.2}$$

where t is the critical path delay in the N branch of the conditional, c is the delay between the condition evaluation node and the preceding memory operation, RD is the RowDecode delay, and P is the Precharge delay. The RowDecode and Precharge stages in the N-branch proceed in parallel with the other computation. Hence, Equations 3.1 and 3.2 need to be satisfied in order to ensure that the unnecessary RowDecode and Precharge stages do not form the critical path in the N-branch.

A similar transformation is shown in Figure 3.10(d). In this case, the Y and N branches have accesses to data in different pages. However, we can pre-issue the RowDecode node for one access, say the *Read(x)* node in the Y branch, (this decision can be either arbitrary, or based on profiling information) and compensate in the N-branch with a Precharge node, provided the operation does not delay the *Write(y)* operation. The following equation needs to be satisfied:

$$t' \geq P \tag{3.3}$$

$$c \geq RD \tag{3.4}$$

where t' is the critical path delay between the conditional and the first memory access in N branch of the conditional, and c, RD, and P are as in Equations 3.1 and 3.2.

3.1.3.4 Loop Transformations. As seen in examples such as that in Figure 3.4, the page mode operation can be efficiently utilized to speed up the schedules for behavioral loops. The CDFG for the behavior has to be transformed to reflect the page mode operation. Since most of the computation occurs in the innermost loops of nested loop structures, we concentrate on the memory accesses in the innermost loops. We consider optimizations for three different cases:

1. loops accessing a single page per iteration;
2. loops accessing multiple pages per iteration; and
3. loops with disjoint subgraphs in the loop body.

(1) Loops accessing single page per iteration.

If a loop accesses locations from only a single memory page per iteration, e.g., there is only one read operation (e.g., *Read* $a[i]$) per iteration, the page mode operations can be applied directly by restructuring the loop so that it iterates over the array in blocks of P iterations (where P is the page size, in words), so that we have one page mode read cycle for every P iterations.

Figure 3.11(a) shows an example of a loop with a single memory access in one iteration, for which a section of the CDFG is shown in Figure 3.11(c). In order to utilize the page mode operation, we introduce an inner loop in which up to P elements from the same page are accessed. Thus, all accesses in the inner loop are from the same page. The transformation is illustrated in Figure 3.11(b). The transformed CDFG is shown in Figure 3.11(d). Note that the RowDecode and Precharge nodes enclose the inner CDFG loop, forming one complete page mode operation. Thus, one iteration of the outer loop results in one complete page mode operation.

The following cases can be handled using the above transformation, or straightforward extensions thereof: (1) loops with a write or R-M-W operation; (2) array index expressions of the type $a[i \pm k]$, where k is a constant; (3) arrays not aligned to page boundaries; (4) loops with constant stride greater than 1; and (5) loop iterations with more than one access to the same page, e.g., $a[i]$ and $a[i+1]$. The transformation in Figure 3.11 improves performance, but could increase the controller complexity. We observe that the increase in complexity is negligible compared to the performance gain (Section 3.1.3.7).

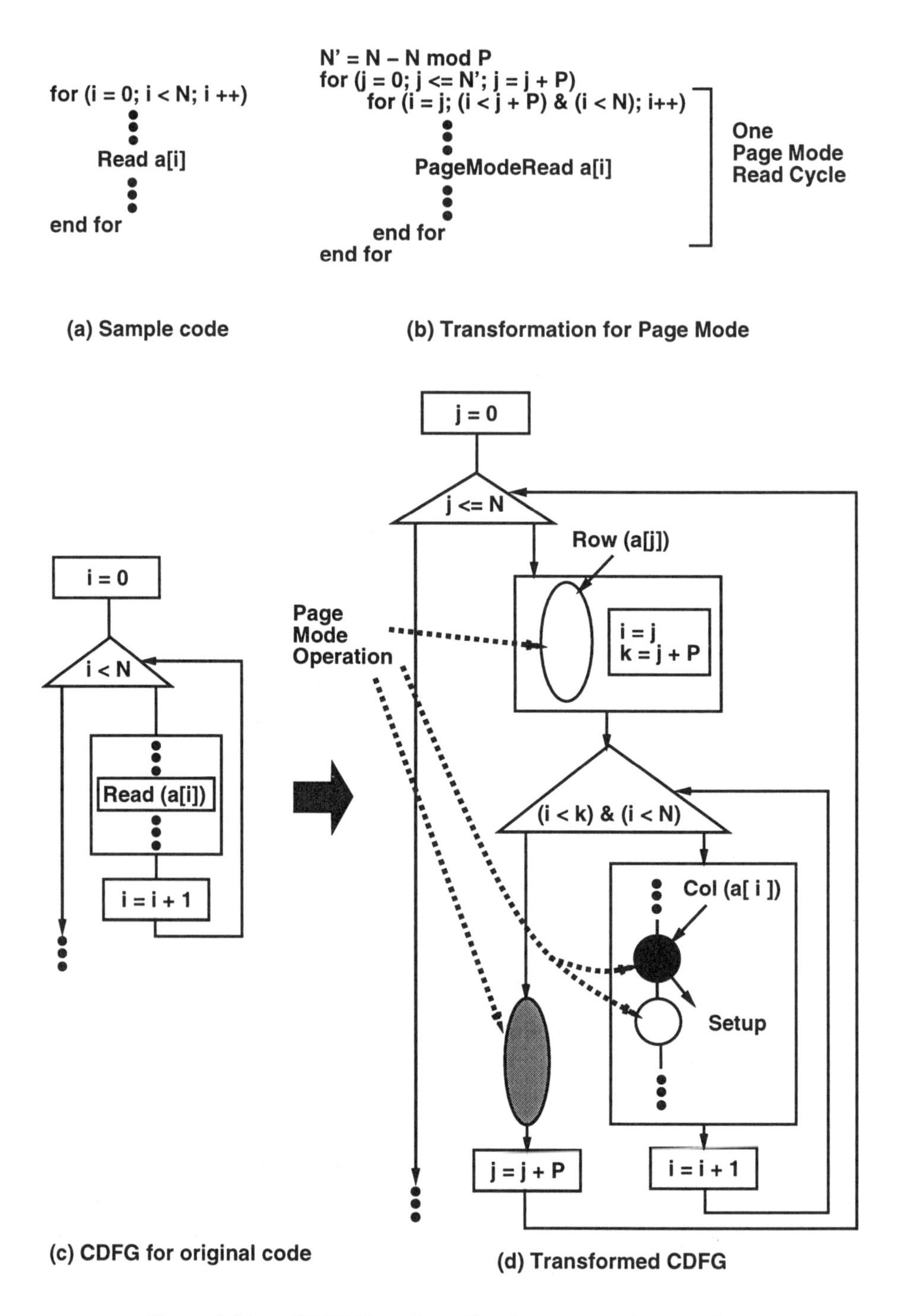

Figure 3.11. CDFG Transformation for page mode operation

(2) Loops accessing multiple pages per iteration.

If more than one array is accessed in one loop iteration, the transformation shown in Figure 3.11 cannot be applied directly, because different arrays usually lie in different memory pages, so that it is no longer possible to use the page mode operation across different iterations. In such cases, we can use a well known transformation, *loop unrolling*, to create the opportunity for utilizing page mode operations. Loop unrolling can result in guaranteed performance improvement when used to exploit page mode operations. Loop unrolling has traditionally been used by compilers to reduce loop overhead, increase instruction-level parallelism, and improve cache locality[BGS94]. In the context of HLS, the data read from off-chip memory has to be stored in registers. Thus, we need to determine the loop unrolling factor based on the size of the available register file.

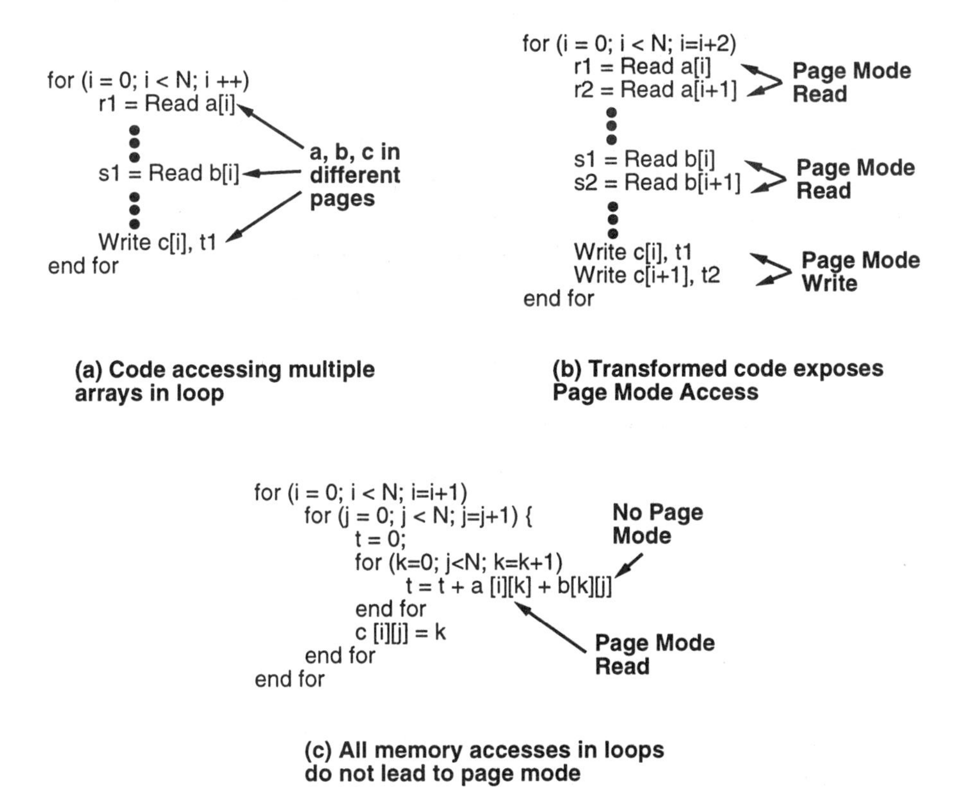

Figure 3.12. Loop unrolling helps exploit page mode operations

Figure 3.12(a) shows an example loop in which three different arrays, a, b, and c are accessed in an iteration. Since the arrays are in different pages, page mode cannot be directly utilized. However, if the loop is unrolled once (Figure 3.12(b) [2]), elements of the same array can be accessed in succession, leading to performance improvement resulting from page mode operation. In Figure 3.12, $r1, s1, t1$, etc. are registers into which memory elements are read in a read operation, and from which memory elements are written in a write operation.

Note that the unrolled loop in Figure 3.12(b) has a higher register requirement. Thus, the unrolling factor is constrained by the maximum number of on-chip registers allowed. Suppose we are given a register file of R registers. We first do an initial scheduling of the loop body (basic block) using *list scheduling*[GDLW92], a well known scheduling algorithm, to determine the number of registers, r, required for one iteration. [3] Let m be the number of memory accesses in the loop body. Note that all memory accesses will not necessarily result in page mode accesses after unrolling. For example, in the code shown in Figure 3.12(c), the access to $a[i][k]$ results in a page mode read operation, but the access to $b[k][j]$ does not, since consecutively accessed elements of b are not in the same memory page. Thus, out of the m memory access in the inner loop body, let m' be the number of accesses that result in page mode accesses. If the loop is unrolled i times, we need $(m' \cdot i)$ registers to store the $(m' \cdot i)$ values. However, the remaining $(r - m')$ registers, which are used in the loop body to store temporary variables and non-page mode memory accesses, need not be duplicated, since they can be reused in the different iterations that constitute the unrolled loop. Thus, if the total number of registers allowed in the register file is R, we must have:

$$m' \cdot i + (r - m') \leq R \tag{3.5}$$

i.e., the loop unrolling factor, i, is given by:

$$i \leq \frac{R - r + m'}{m'} \tag{3.6}$$

or

$$i = \left\lfloor \frac{R - r + m'}{m'} \right\rfloor \tag{3.7}$$

In the example in Figure 3.12(c), the b array can be transposed to lead to consecutive access in the inner loop (in Section 3.1.3.7, we analyze an optimized version of the code in Figure 3.12(c)). Similarly, in the example of Figure 3.12(a), physically interleaving the three arrays a, b, and c in memory allows page mode to be utilized without unrolling. However, such data reorganizations have to follow a more global analysis, because of

[2] If N is not exactly divisible by the unrolling factor (in this case, 2), then a *residue* loop will have to be inserted. We omit this in our example for the sake of clarity.

[3] The required number of registers is the maximum number of data dependence edges that cross any clock cycle boundary of the schedule.

possibly conflicting requirements in different loops. Data reorganization techniques for improving spatial locality are discussed in [CL95], as well as in Section 3.2. We assume that such data transformations have been performed to the code prior to this step. Note that the unroll factor derived here is somewhat pessimistic – a higher unroll factor could be obtained at the expense of computation time by alternately increasing the unroll factor and performing scheduling and register allocation on the unrolled loop until the register requirements cannot be satisfied. A detailed treatment of the interaction between register allocation and loop unrolling is found in [KNDK96].

(3) Loops with disjoint subgraphs in body.

A loop body is said to consist of disjoint subgraphs if the DFG representing the body can be divided into more than one subgraph with no data dependence across their memory operations (i.e., no data dependence edges from one subgraph to the other). In such a case, each subgraph with at least one memory access in it, can be split into a different loop, in order to better utilize page mode memory operation.

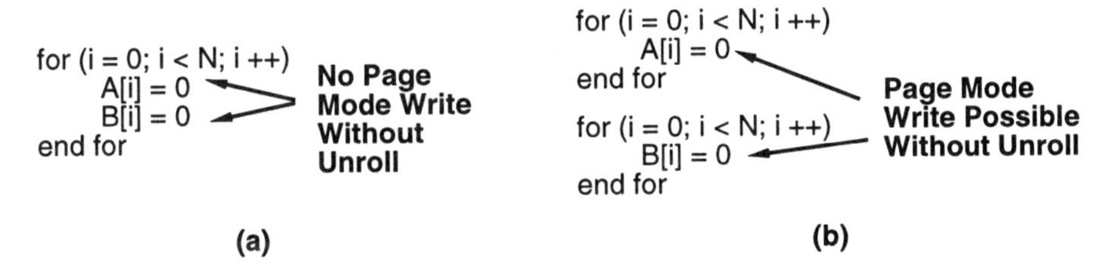

Figure 3.13. (a) Loop with disjoint subgraphs in body – there is no data dependence between the 2 statements (b) Split loop – page mode write can now be applied to the individual loops

An example loop with disjoint subgraphs in its body is shown in Figure 3.13(a). The two statements in the loop body have no data dependence. We can split the loop into two loops, as shown in Figure 3.13(b), so that page mode write can now be applied to the individual loops without unrolling. Two points are worth noting here:

- Page mode operations without loop unrolling are preferable, since unrolling increases the register requirements, and also limits the number of memory accesses in one page mode cycle, due to register file size constraints.
- Loop splitting increases the total number of operations slightly due to the repeated index increments and comparisons, but the overall time saved by exploiting page mode operations is much larger. In the example in Figure 3.13(b), the splitting leads to 2 extra (usually single-cycled) operations: (1) the loop index increment ($i = i + 1$), and (2) the comparison ($i < n$) in each loop iteration. However, this

overhead is more than offset by the fact that page mode operations require only 2 cycles per memory access in the steady state, while individual read and write cycles require 7 cycles. Further, there is the possibility that, as in Figure 3.13(b), the 2 extra operations could be scheduled in parallel with the memory access, leading to no performance overhead.

3.1.3.5 The CDFG Transformation Algorithm. Algorithm *TransformCDFG* in Figure 3.14 outlines the sequence of steps for transforming the CDFG of the behavior into an optimized form, so that a scheduling algorithm, such as list scheduling, can be invoked efficiently on the transformed CDFG. The inputs to the algorithm are the CDFG for the behavior, the page size P of the memory, and the maximum register file size allowed, R. We first cluster the scalar variables into groups of size P (Section 3.1.3.1). In Step 2, we perform the reordering optimization outlined in Section 3.1.3.2. Step 3 performs the hoisting transformation (Section 3.1.3.3) to the conditional nodes in the CDFG, if applicable. In Step 4, we perform the loop splitting transformation (Section 3.1.3.4) on loops with disjoint subgraphs in their bodies. For each innermost loop in the resulting CDFG, we perform the page mode transformations of Section 3.1.3.4 if there are accesses to only a single page in the loop body; otherwise, we do a preliminary scheduling of the loop body using the list scheduling algorithm to determine the total number of registers used, r, and the number of registers involved in page mode operations, m'. We then unroll the loop $\lfloor \frac{R-r+m'}{m'} \rfloor$ times and reorder the accesses that can now be treated as page mode accesses. We can now invoke a scheduling algorithm such as list scheduling on the entire transformed CDFG.

Although Step 1 has a complexity $O(Pm^2)$, where m is the number of scalars, this step is fast in practice, because the number of scalars mapped into off-chip memories is small in comparison to the page size. Steps 3 and 4 are both $O(n)$ routines, where n is the number of nodes in the CDFG. The list scheduling algorithm in Step 5 requires $O(n^2)$ time for each loop nesting. Thus, Step 5 requires $O(Bn^2)$, where B is the number of loop nests in the behavior. Thus, the overall complexity of the algorithm is $O(Bn^2)$, the complexity of the dominant Step 5.

3.1.3.6 DRAM Refresh. An important property of dynamic RAM is that the internal data bits need to be refreshed at regular intervals. The refresh intervals of typical DRAMs are fairly large: for instance, the IBM11T1645LP has a refresh interval of 128 ms. The time taken to perform a single *Refresh cycle* is comparable to the duration of any other access, such as *Read cycle*. In the IBM11T1645LP, the refresh cycle requires roughly 80 ns (compared to 105 ns for read cycle).

For proper functioning of the DRAM-based system, the refresh cycles have to be properly accounted for by the controller in order to preserve the correct data in the

Algorithm *TransformCDFG*
Input: G – CDFG of Behavior; R – Register File Size;
P – Memory Page Size
Output: Transformed CDFG

1. Cluster scalar variables into groups of size P, and assign memory addresses.
2. **for** each basic block B in G
   Reorder operations to exploit R-M-W mode and page mode operations in B.
3. **for** each conditional node in G
   Perform *hoist* transformation, if applicable.
4. **for** each innermost loop L in G
   Perform Loop Splitting transformation, if applicable.
5. **for** each loop L' in updated G
   if single page accessed in one iteration
      Perform loop restructuring for page mode
   else
      Perform List Scheduling on loop body.
      Let r = total no. of registers required for this schedule.
      Let m' = no. of registers involved in page mode access
      Unroll loop $\lfloor \frac{R-r+m'}{m'} \rfloor$ times.
      Reorder memory accesses in unrolled loop to exploit page mode.
   end if

/ Scheduling can now be performed on the transformed CDFG. */*
end Algorithm

Figure 3.14. Algorithm for incorporating memory optimization transformations into the CDFG for scheduling

DRAM. This is usually done with a *Refresh Counter*, which counts up to the number of DRAM rows. The value of the counter during a refresh cycle gives the row number that is refreshed (assuming one row is refreshed in one refresh cycle). After every refresh cycle, the counter is incremented, in preparation for refreshing the next row. Thus, the controller needs to ensure that the entire sweep through all the DRAM rows occurs within the refresh interval. The refresh can be performed as a post-processing step after a schedule for the input behavior is generated. The schedule is adjusted to incorporate the refresh cycles, ensuring that they do not overlap in time with any other memory operation, since the DRAM is unavailable for Reads and Writes during the refresh.

We now need to determine the number of refresh cycles to incorporate into the schedule in order to ensure that all DRAM rows are refreshed within the refresh interval. Let the refresh interval of the DRAM be R; the number of DRAM rows be n; and the length of the generated schedule be S. In our example memory chip, we have: $R = 128$ ms, $n = 1024$. Since n rows need to be refreshed in time R, the refresh interval can be satisfied if we allow a maximum time $T = \frac{R}{n}$ between refreshes to successive rows. Hence, the number of refresh cycles (r) we need to introduce into the schedule of length S is:

$$r = \left\lceil \frac{S}{T} \right\rceil = \left\lceil S \cdot \frac{n}{R} \right\rceil \tag{3.8}$$

In the example above, for a schedule of length $S = 1024$ ns, we have: $r = \lceil 1024 \cdot \frac{1024}{128 \cdot 10^6} \rceil = 1$. That is, we only need to introduce one refresh cycle into the schedule. In fact, for all schedules less than $R/n = 128000$ ns, we need only one refresh cycle. For an example schedule of length $S = 1$ ms, we have: $r = \lceil 1 \cdot 10^6 \cdot \frac{1024}{128 \cdot 10^6} \rceil = 8$ refresh cycles. Thus, we see that the performance overhead due to refresh requirements is quite modest. Further, this is not an overhead due to our approach because the refresh circuitry is essential for correct DRAM operation, no matter what synthesis technique is used.

3.1.3.7 Experiments. The optimizations for efficiently utilizing advanced memory features were tested on benchmark examples from the digital signal processing and scientific computing domains, all of which share the common characteristic that they process large data arrays. We present a summary of these experimental results in this section.

Characteristics of Benchmark Examples

Column 1 of Table 3.2 shows the list of benchmark examples on which the experiments were performed. The *Beamformer, DHRC, Lowpass,* and *SOR* examples are extracted from the High-level Synthesis Design Repository[PD95a]. *Beamformer* is a DSP application involving temporal alignment and summation of digitized signals from an N-element radar antenna array. *DHRC* (Differential Heat Release Computation) is an algorithm modeling the heat release in a combustion engine [CS93]. *Lowpass* is an image processing application that applies a low-pass filter to an image. *SOR* (Successive Over-Relaxation) is an algorithm used in evaluating partial differentiation equations. *Dequant* is the de-quantization routine in the MPEG decoder application [Gal91]. *IDCT* (Inverse Discrete Cosine Transform), *LeafComp*, and *LeafPlus* are modules from the MPEG decoder application[Gal91]. *Madd* and *MMult* are matrix addition and multiplication routines respectively. Column 2 shows the number of basic blocks in each benchmark.

Table 3.2. Memory optimizations applied to benchmarks

		Memory Modes				Optimizations			
Benchmark	B	rmw	pr	pw	prmw	C	R	H	L
Beam	10	Y	Y	Y	N	N	Y	N	Y
Dequant	5	N	Y	Y	N	N	N	Y	Y
Dhrc	2	N	Y	N	N	N	N	N	N
Idct	13	N	Y	Y	N	N	N	N	Y
LeafComp	7	N	Y	Y	N	N	N	Y	Y
LeafPlus	5	N	Y	Y	N	N	N	Y	Y
Lowpass	4	Y	Y	N	N	N	Y	N	N
Madd	4	N	Y	Y	N	N	N	N	Y
MMult	6	N	Y	N	Y	N	N	N	Y
SOR	4	Y	Y	N	N	N	Y	N	N

Table 3.2 shows the memory modes utilized by the memory accesses in the various benchmark examples, and the applicable optimizations. Columns 3, 4, 5, and 6 indicate (with letter 'Y') examples that had memory access patterns for which the Read-Modify-Write (rmw) mode, page mode read (pr), page mode write (pw), and page mode read-modify-write (prmw) respectively, were applied. Columns 7, 8, 9, and 10 show the CDFG transformation techniques – *Clustering*(C), *Reordering*(R), *Hoisting*(H), and *Loop Transformations*(L) – that were applied to each example. Note that the scalar clustering technique could not be applied to any of the examples, as the number of scalars in the examples was too small, and could be stored in the register file we used.

Illustrative Example: Matrix Multiplication

Matrix Multiplication is an important procedure used in not only numerical algorithms, but also in DSP applications such as Discrete Cosine Transform (DCT). Figure 3.15 shows the code for the matrix multiplication procedure using the *ikj*-loop ordering, which is known to have better spatial locality than the standard *ijk*-version [LRW91]. Note that page mode write can be applied to the initialization loop with the transformation shown in Section 3.1.3.4, since only a single array is involved. No loop unrolling is necessary in this case. Further, as pointed out in Figure 3.15, the reading and writing of $z[i][j]$ can be combined into a single R-M-W operation. The operations in the innermost loop are ordered, so that the multiplication "$r * y[k][j]$" is completed before $z[i][j]$ is read. Similarly, the innermost loop (index j) can be unrolled to exploit both page mode read for array y, and page mode R-M-W for array z.

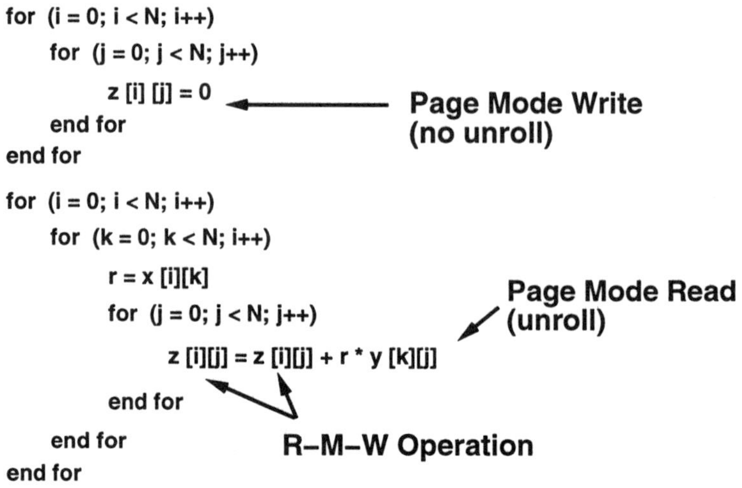

```
for (i = 0; i < N; i++)
    for (j = 0; j < N; j++)
        z [i] [j] = 0         <------ Page Mode Write
    end for                           (no unroll)
end for

for (i = 0; i < N; i++)
    for (k = 0; k < N; i++)
        r = x [i][k]
        for (j = 0; j < N; j++)               Page Mode Read
            z [i][j] = z [i][j] + r * y [k][j]  (unroll)
        end for
    end for                   R-M-W Operation
end for
```

Figure 3.15. Matrix Multiplication Example

We compare the execution time due the schedules generated by 3 techniques:

Coarse-grain – the traditional HLS approach, where memory access operation is treated as a multicycled operation, with the *read* data being available only at the end of the read operation, and the *write* data being required to be available at the beginning of the write operation. List scheduling is used to generate the schedule.

Fine-grain – more refined memory access modes (e.g., the template strategy of [LKMM95]), with each memory access operation being treated as an independent 3-stage operation, using the models shown in Figures 3.3(b) (for read operation) and 3.6(a) (for write operation). List scheduling is used to generate the schedule.

Optimized – the scheduling algorithm described in Figure 3.1.3.5, which incorporates the CDFG transformation techniques outlined in Section 3.1.3, making efficient utilization of the various memory modes.

For the matrix multiplication example, the *Coarse-grain* technique resulted in 4.628×10^7 cycles; the *Fine-grain* technique resulted in 4.419×10^7 cycles; and the *Optimized* technique resulted in 2.217×10^7 cycles.

Summary of Results

Table 3.3 shows the values of the various applicable parameters used in our experiments, using the timing characteristics of the IBM11T1645LP memory chip. The characteristics of other commercial DRAM chips are similar, so the delays used are representative of typical DRAMs.

Table 3.3. Parameters used in Experiments

Parameter	Value
Operator Delay (+, −, >, Shift)	1 cycle
Operator Delay (*)	2 cycles
Operator Delay (/)	4 cycles
Operator Delay (RowDecode, Precharge)	3 cycles
Operator Delay (ColDecode, Setup)	1 cycle
Memory Page Size	256 words
Register File Size	16 words
No. of ALU, Multiplier, Divider	1
No. of Read Ports in Register File	2
No. of Write Ports in Register File	1

Figure 3.16 summarizes the experimental results for the benchmark examples on which we tested our technique. Figure 3.16(a) shows the number of clock cycles required by the coarse-grain technique for each benchmark. On the x-axis of Figure 3.16(b), we show the benchmark examples of Table 3.2, and on the y-axis, we show the total execution times for the schedules generated by the three techniques: Coarse-grain, Fine-grain, and Optimized, normalized to the cycle count of Coarse-grain, which is taken as 100. The normalization is necessitated by the widely varying cycle counts of the examples, which would make the comparison difficult if plotted on the same graph.

We observe that Fine-grain always results in faster schedules than Coarse-grain, because the Fine-grain technique incorporates two important observations about memory accesses: (1) in the read mode, the data is available before the entire memory read cycle completes; and (2) in the write mode, the actual data need not be available at the beginning of the memory write cycle. Both these observations lead to other operations being scheduled in parallel with the Precharge and RowDecode nodes, leading to faster schedules. These observations are also incorporated into the Optimized technique, which utilizes several additional optimizations.

The Optimized technique results in the fastest schedules, as seen from Figure 3.16. On an average, Optimized achieves a performance improvement of 45.2% over Coarse-grain, and an improvement of 40.8% over the Fine-grain technique. The unrolling factor for the examples varied from 0 to 7. The performance improvement in the case of Optimized is the consequence of efficient utilization of the memory features, such as read-modify-write, page mode read, page mode write, and page mode read-modify-write, all of which lead to reductions in memory access time. These modes are not

Benchmark	Cycle Count (Coarse-Grain)
Beam	1,251,844
Dequ	4,226
Dhrc	6,784
Idct	63,521,284
LeafC	2,562
LeafP	3,074
LowP	1,159,202
Madd	360,706
MMult	46,285,058
SOR	690,860

(a)

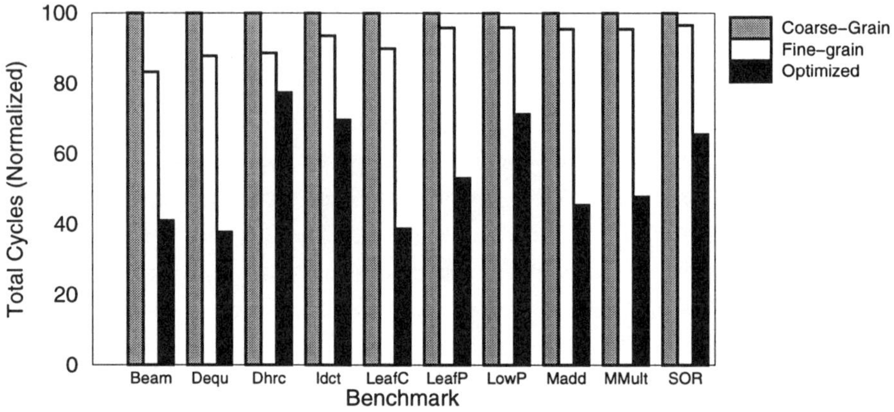

(b)

Figure 3.16. (a) Cycle count for Coarse-Grain (b) Summary of Results

exploited by the Coarse-grain and Fine-grain techniques, which treat each memory access independent of the others.

Controller Complexity

For some memory access modes, the performance improvement for the Optimized technique above could incur a small overhead in controller area due to increased complexity, as compared to the controllers required for the Coarse-grain and Fine-grain schedules. In other modes, the controller complexity might actually reduce. The following cases are worth noting.

1. In the Read-Modify-Write operation, the number of states required by the Optimized technique is less than that of Coarse-grain and Fine-grain. Also, the number of

operations initiated by the controller is reduced, since, both RAS and CAS are asserted (and de-asserted) only once in the R-M-W cycle. However, RAS and CAS have to be asserted (and de-asserted) twice in both Coarse-grain and Fine-grain, since the read and write are treated as separate memory operations. Thus, when R-M-W operations are used, the controllers generated by Optimized are smaller.

2. When page mode operations are used within a basic block, e.g., combining accesses to two variables a and b into a single page mode access, the number of states, as well as the number of RAS and CAS assertions (and de-assertions) are reduced in the Optimized technique, as above, leading to smaller controllers.

3. When page mode operations are a result of loop transformations, the Optimized technique introduces an extra state (Setup). Also, some new operations are introduced, e.g., in Figure 3.11(d), the assignments "$k = j + P$" and "$j = j + P$"; and the comparisons "$(i < k) \& (i < N)$". This could increase the area of the controller.

Table 3.4. Area of generated controllers

	Controller Area (in literals)		
Benchmark	Coarse-grain	Fine-grain	Optimized
Beamformer	1149	1339	1543
Dequant	780	966	1064
Dhrc	1102	1235	1146
Idct	1273	1355	1672
LeafComp	737	730	1078
LeafPlus	478	575	559
Lowpass	1237	1386	1097
Madd	564	443	468
MMult	676	720	755
SOR	1506	1916	1458

We synthesized the controllers for each of the schedules generated by the three techniques on the examples of Table 3.2, using the *misII* synthesis package[BRSV87]. Table 3.4 shows the complexity of the controllers in terms of literal count, which is a well-known technology-independent metric for comparing FSM area. Note that in many cases (DHRC, SOR, Lowpass, and LeafPlus), the controller area actually decreased; these occurred when cases 1 and 2 above applied. On an average, the controllers generated from the Optimized schedules have an area only 14.9% larger than those generated from Coarse-grain, and 10.3% larger than those generated from Fine-grain. Given the significant improvement in performance with our optimizations, the control overhead is justifiable. Note that there is no address computation overhead due to the transformations.

In summary, our memory optimization techniques result in significant performance improvements in all the benchmark examples, all of which are memory-intensive. The performance improvements are due to the CDFG transformation techniques we employ to exploit efficient memory access modes. The impact of the transformations on the controller area is minimal – in fact the resulting controllers were smaller in some cases.

3.2 POWER OPTIMIZATIONS IN MEMORY ACCESSES

In Section 3.1, we discussed optimization techniques for exploiting locality in memory accesses, which led to significant performance improvements. Locality of reference also plays a major role in reducing power consumption. As explained in Section 3.1.1, consecutive memory accesses from the same page lead to a performance gain because the precharging and row decoding phases can be omitted for the second access. This also results in a corresponding decrease in power consumption, because omission of the precharging and row decoding phases represents reduced switching activity in the circuit.

It is a well known fact that signal transitions, as represented by switchings in logic-level bit values, are the primary source of power consumption in CMOS circuits. Thus, most attempts at low power design of CMOS at the logic or behavioral levels usually include an effort at minimizing the number of signal transitions. Regularity in memory accesses can be exploited by deriving address access patterns from the behavior and using these to reduce the number of transitions on the memory address bus.

The reduction of transition counts on the memory address bus has two important implications. First, it leads to a reduction in the switched capacitive load on the off-chip drivers of the address bus, which results in significant power savings, since off-chip capacitances are three orders of magnitude larger than typical on-chip capacitances [Bak88]. Second, reduced activity in the address bus also leads to reduced activity in the memory address buffers and decoding circuitry. Studies have shown that power dissipation in the address decoder and address buffers of typical memory chips constitute a significant portion of the power consumed (up to 50%) in the memory chip [WCF+94]. This implies that design techniques leading to decrease in power dissipation in this part of the memory will make a significant impact on the overall power dissipation of the application.

In this section, we focus on the reduction of power dissipation in off-chip drivers and in the memory's decoding logic by reducing the number of transitions on the memory address bus. The technique is an extension of the compiler techniques for reducing the number of memory accesses – we exploit regularity and spatial locality in the memory accesses and determine the mapping of behavioral array references to physical memory locations to minimize address bus transitions. The technique can be applied in conjunction with other approaches ([WCF+94, SB95]) to obtain further power minimization at the behavioral level.

We study the address bus power reduction problem for three different memory configurations:

1. single memory (single-port),

2. multiple memories (single-port), and

3. single multiport memory.

3.2.1 Mapping Arrays into a Single Memory

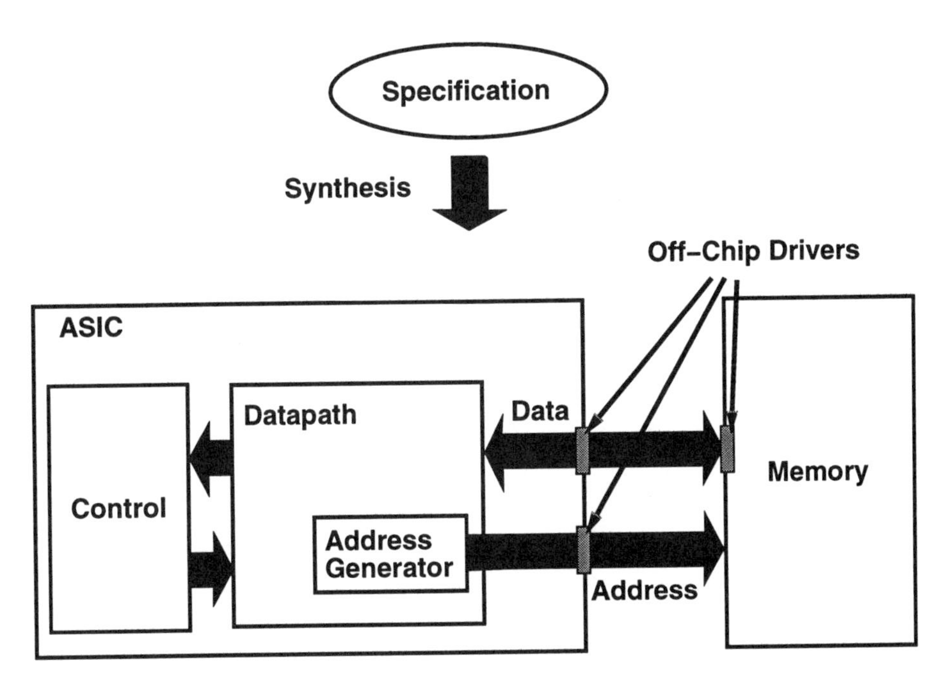

Figure 3.17. Synthesis model of a memory-intensive system

3.2.1.1 Problem Definition. Figure 3.17 shows a more detailed view of the highlighted blocks in Figure 3.1 for a typical memory-intensive system, synthesized into an ASIC (consisting of datapath and control blocks) and an off-chip memory. Memory accesses result in power dissipation in both the ASIC and memory. Power dissipation in the ASIC is the sum of power dissipation in its constituents – the datapath, the control, and the off-chip drivers. A reduction in the transition count of the address bus also entails power savings in the off-chip drivers of the ASIC.

Power consumption during memory accesses within a memory can be roughly classified into two categories – *data-related* and *address-related*. *Data-related* power is the fraction of memory power dissipated during data transfers between the data bus and the memory location where the data resides, and the dissipation in the refresh circuitry (if any) of the memory. *Address-related* power represents the fraction contributed by the address buffers and decoders. Since data-related power is dependent on the specific input data to the application, we have no control over minimizing power without knowledge of the data profile. However, we always have an opportunity to minimize address-related power, since the application's memory access sequence patterns are known at the specification stage. These memory access sequences can be exploited to guarantee a reduction in the memory address bus transitions, thereby leading to guaranteed power reduction during memory accesses.

Given a behavioral description to be synthesized, our task is to determine the assignment of arrays in the specification to physical addresses in memory. For simplification of our discussion, we assume that every array element occupies one memory word, although this assumption is not necessary in the analysis.

3.2.1.2 Array Mapping Strategy. We first assign the arrays in the specification to locations in a *logical memory*, and employ a Gray Code Converter (GCC) in the *address generator* (Figure 3.17) to map logical memory addresses to physical memory locations. The GCC converts a logical address into *Gray code* [Koh78] thereby ensuring that access of consecutive logical memory addresses results in the transition of exactly one bit on the memory address bus. The GCC helps bring down the address bus transition count to a minimum, but the memory mapping strategies we discuss are valid techniques for ensuring reduction in the transition count even in the absence of the GCC from the architecture.

We concentrate on memory mapping strategies for two dimensional arrays. We consider three strategies in our attempt to find an effective mapping of arrays into memory:

Row-Major. A simple way of mapping a logical array to physical memory (unless the order is imposed by the language) is to store the elements in row-major form [ASU93], that is, the elements of the first row are placed in consecutive memory locations in order of increasing column index. This is followed by the elements of the second row in the same order, and so on.

Column-Major. In column-major mapping [ASU93], the elements of the two dimensional array are stored column-by-column.

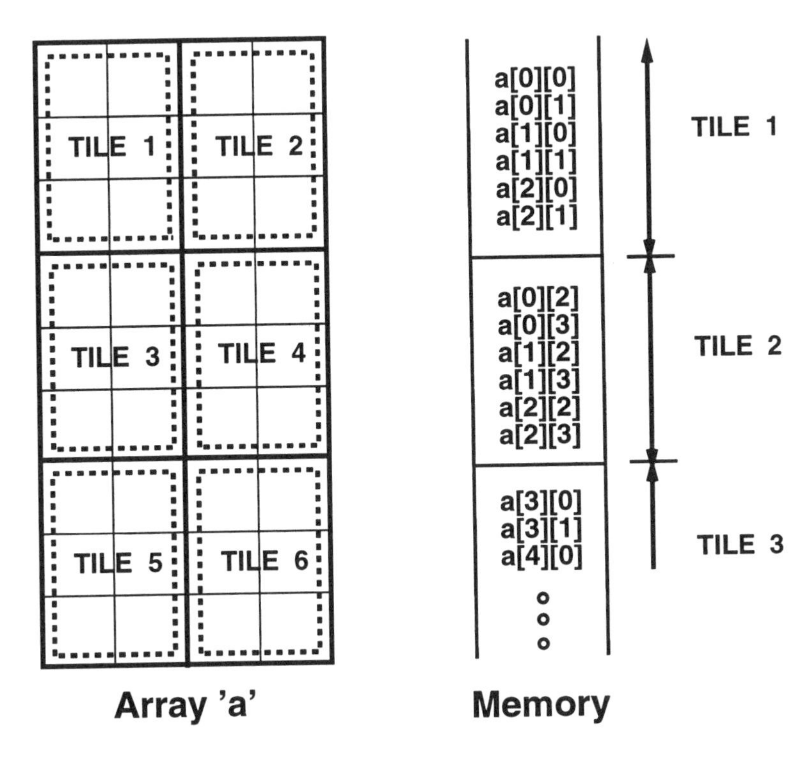

Figure 3.18. Tile-based mapping of a 4×9 array – with tiles of size 2×3

Tile-Based. This mapping style first partitions the behavioral array into tiles of smaller rectangles. Here, the elements of a row (or column) of tiles are stored in consecutive memory locations, with the tiles themselves being stored in either row or column major format (Fig. 3.18).

Similar ideas have been used in the context of compilers, where cache reuse is improved by dividing a loop's iteration space into tiles and transforming the loop nesting structure to iterate over the tiles (Chapter 4). Block decomposition of arrays in a multiprocessor environment [BGS94] is also based on a similar concept.

Figure 3.19(a) shows a simplified version of the code kernel for a Successive Over Relaxation (SOR) algorithm [PTVF92], often used in the domain of Image Processing applications. The plus-shaped contour in Figure 3.19(b) shows the basic access pattern of the elements of array u in the inner loop. Figure 3.19(c) shows that this shape of accesses moves across by two columns to the right as we iterate through the inner loop. The second iteration of the outer loop causes the first pattern of the previous loop to move vertically down by one row. The remaining accesses of the inner loop are as before.

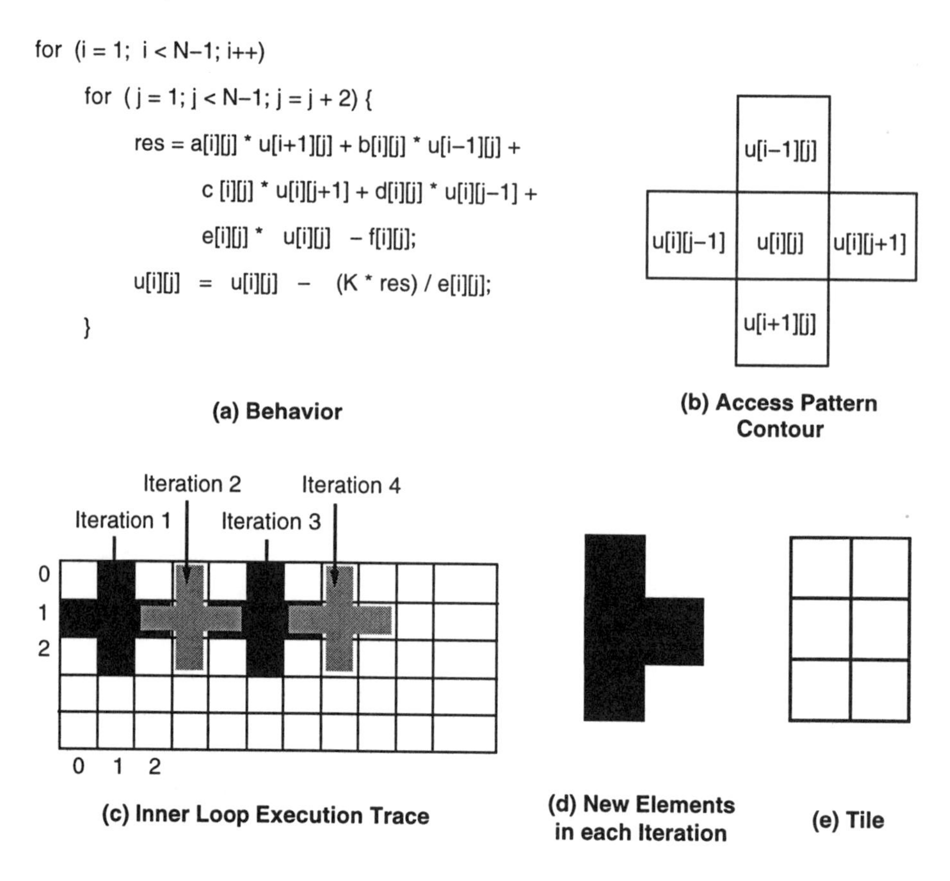

Figure 3.19. Example (SOR) – Access Pattern of Array 'u'

The one element that is common to successive patterns can be stored in a register instead of being accessed from memory again in the next iteration [KND94], hence the effective access pattern for this example is as shown in Figure 3.19(d). This pattern determines the dimensions of the tile to be used in the tile-based mapping scheme. The smallest rectangle that encloses this repeating shape forms the tile we use (Figure 3.19(e)).

We now examine the effect on the number of address bit transitions in the three mapping styles for the SOR example of Figure 3.19. This analysis helps determine the mapping style that minimizes the total transition count.

We use the observation that, in general, there is a low Hamming distance between elements in the same tile or adjacent tiles, and a possibly high Hamming distance between elements in distant tiles. We define a *maximal transition* as occurring when two logical addresses with a large difference are accessed in succession. This is because two numbers with a large difference are, on an average, likely to have a large Hamming Distance. A *minimal transition* occurs when this difference is small. In this case, *large* means comparable to the dimension of the array. For example, we treat all consecutive accesses to elements in the same tile (or adjacent tiles) as minimal transitions.

The above classification of maximal and minimal transitions is related to the DRAM organization (Section 3.1.1). If the arrays involved are large, and are stored in different memory pages, a maximal transition refers to memory accesses from two different pages, and a minimal transition refers to consecutive memory accesses from the same page. If the arrays are small, and fit into the same DRAM page, the transitions refer to the switching activity on the column address (since the row address is now identical for both accesses).

In [PD95b] it is shown that if column-major mapping is used when the tile has n columns, there will be at least n maximal transitions in every inner loop iteration (assuming at least one element from every column is accessed). Similarly, row-major mapping would entail n maximal transitions in an n-row tile. In comparison, tile-based mapping leads to a maximum of 2 maximal transitions, independent of the tile size, provided sufficient number of registers are available. [4] We thus choose to use tile-based mapping when the tile has both dimensions greater than 2. Otherwise, we use row- or column-major as appropriate. The above observation leads to the heuristic *SelectMapping* in Figure 3.20 for selecting a mapping scheme for a 2-level nested loop, involving a two-dimensional array.

The reason for the condition "*j-increment in outer loop* = L_j" is that, if the increment of the outer loop is the same as the tile dimension, then every iteration of the inner loop will access elements from a single tile. This obviously makes tile-based mapping a preferable style, since all the transitions will be minimal in every iteration. If condition (1) fails, the extracted tile has dimensions N $\times$ 2, N $\times$ 1, 2 $\times$ N or 1 $\times$ N. Condition (2) ensures that for the first two cases, column-major mapping is selected, and row-major mapping is selected for the other two. If $L_j = L_i$, we arbitrarily select row-major mapping.

3.2.1.3 Implementation Overhead. In a typical implementation of the address generator, we have the following *index expressions* for generating the logical address for an arbitrary element $u[i][j]$:

Row-major – $A + (i \times MAX) + j$

Column-major – $A + (j \times MAX) + i$

[4] If registers are insufficient, there will be more than 2 maximal transitions, but the sequence can always be ordered to perform as well, or better than the row-major case (in row-major, a maximal transition occurs every time the row changes).

Heuristic *SelectMapping*
Input: Memory References to Array in Loop
Output: Storage Scheme for Array

Let the inner loop index be i and outer loop index be j
Extract basic repeating shape from the access patterns
Determine the enclosing rectangle R.
Let i-dimension of R be L_i and j-dimension be L_j
if j-increment in outer loop = L_j **or** $\min(L_i, L_j) > 2$ **then** (1)
use *tile-based* mapping
else
if $L_j > L_i$ (2)
then use *column-major* mapping
else
use *row-major* mapping
end if
end if
end Heuristic

Figure 3.20. Selection of Memory Mapping Strategy

Tile-based – $A + (i \times MAX) + (N \times j) - (i \bmod N) \times (MAX - 1)$, [5]
where the tile dimensions are $M \times N$ and A is the start address of array u of dimensions $MAX \times MAX$ (i and j range from 0 to $MAX - 1$).

Note that the expressions above need to be evaluated only if an arbitrary element $u[i][j]$ is accessed *outside of any regular loop structure.* However, since most of the computation (and hence, power dissipation) occurs within loops, we can use techniques similar to *value numbering* [Muc97], *strength reduction* and *induction variable elimination* [ASU93] to minimize the actual number of operations. This means, for example, that the value $(i \times MAX)$ is never computed by a *multiplication* operation, but is implemented by *addition* of MAX to a running counter every time i is incremented in a loop.

We now illustrate a possible hardware implementation of the inner loop access sequences for array u in the SOR example of Figure 3.19 using row-major and tile-

[5] This expression is independent of M – we have assumed that the elements within a tile are stored column-wise, while the tiles themselves are stored row-wise; in other (symmetrical) cases, it will be independent of N.

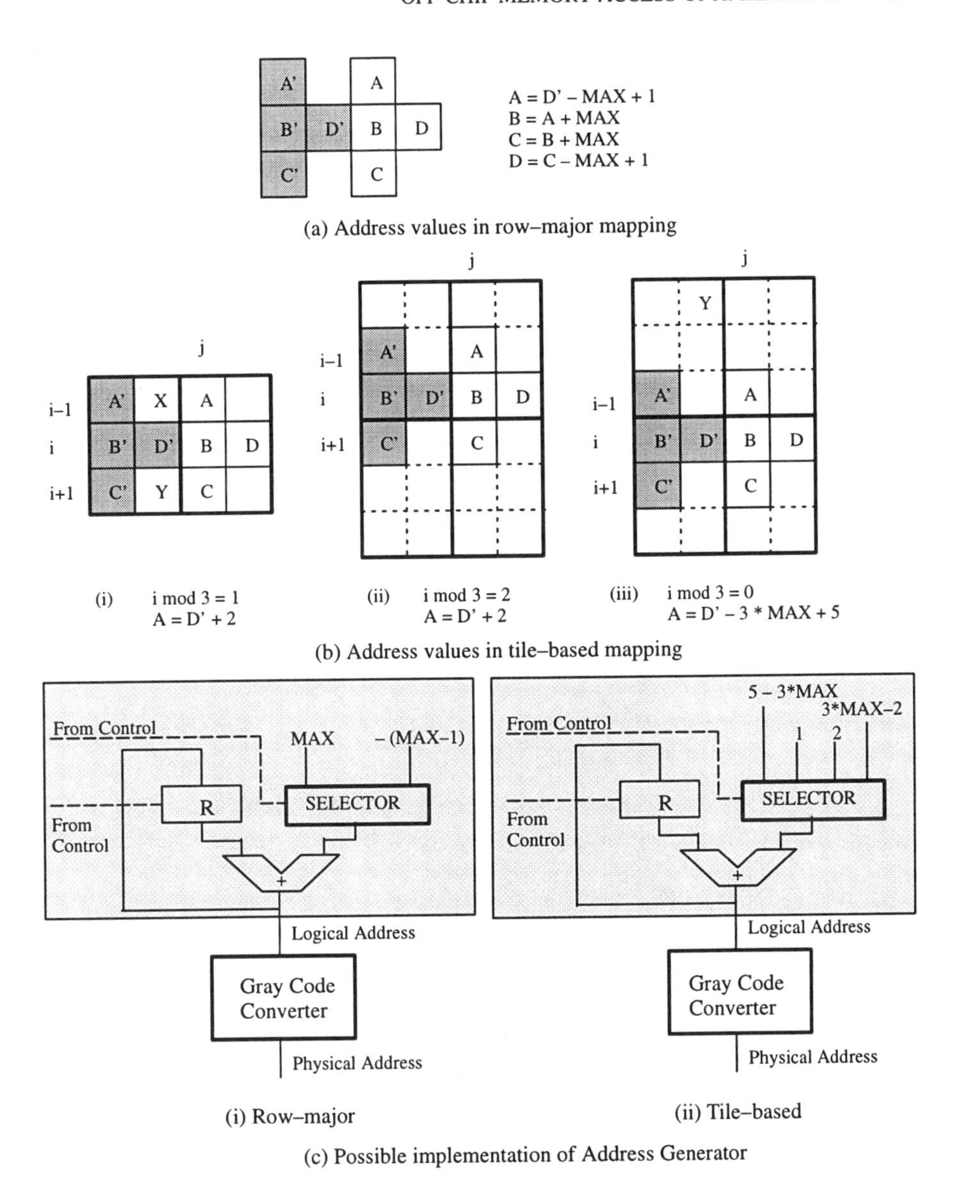

Figure 3.21. Implementation of Address Generator

based mapping. Figure 3.21(a) shows the elements accessed in two consecutive iterations of the inner loop. The four elements of array u accessed in the current loop iteration are at logical addresses A, B, C and D. The four elements accessed in the previous iteration are at addresses A', B', C' and D'. For row-major mapping, it is easy to see that D', A, B, C and D are related through the equations in Figure 3.21(a).

Each address is generated from its preceding address, and we need to maintain only the most current address in a register. [6] We need only one addition operation to generate the address for every memory access in the loop ($(MAX - 1)$ is a constant known at compile time).

For tile-based mapping, Figure 3.21(b) shows the computation of location A in terms of D'. We have the following equations:

$$A = \begin{cases} D' + 2 & \text{when } i \bmod 3 = 1, 2 \\ D' + (5 - 3 \times MAX) & \text{when } i \bmod 3 = 0 \end{cases}$$

$$B = \begin{cases} A + 1 & \text{when } i \bmod 3 = 1, 2 \\ A + (3 \times MAX - 2) & \text{when } i \bmod 3 = 0 \end{cases}$$

$$C = \begin{cases} B + 1 & \text{when } i \bmod 3 = 0, 1 \\ B + (3 \times MAX - 2) & \text{when } i \bmod 3 = 2 \end{cases}$$

$$D = \begin{cases} C + 2 & \text{when } i \bmod 3 = 0, 1 \\ C + (5 - 3 \times MAX) & \text{when } i \bmod 3 = 2 \end{cases}$$

We present below a computation of the maximum effective transition count overhead and savings incurred in various parts of the system due to the mapping schemes.

Overhead in Gray Code Converter

In general, an n-bit wide Gray code converter (GCC) is implemented with $n - 1$ XOR gates. The MSB of the input becomes the MSB of the output, while the rest of the output bits are generated by XORing adjacent bits of the input [Koh78]. A typical implementation of a 2-input CMOS XOR gate uses 8 transistors [WE85]. On examining this circuit, we notice that in an XOR gate, there can be a maximum equivalent transition count of 6, which occurs when both inputs change [PD95b]. It can be concluded that the number of internal transitions in the GCC is no more than $6N$, where N is the number of transitions on its inputs (assuming the absence of glitches at the input).

Overhead in Selector The implementation of the address generator for the two mapping strategies is shown in Figure 3.21(c). The SELECTOR circuits in the two mapping styles are different. They use information from the controller to select the appropriate (constant) operand for addition. Figure 3.22 shows the structure of the selector

[6] If the value of i and j are needed later, they can be assigned the final value *once, after the loop.*

circuits for row-major and tile-based mapping. In both cases, a two-bit signal from the controller ($a_1 a_0$) indicates which of the four addresses (A, B, C and D) is being accessed. In Figure 3.22(a), the combinational circuit generates the appropriate value for s, so that the correct operand is chosen by the multiplexer. In Figure 3.22(b), the combinational circuit uses, along with the two-bit signal ($a_1 a_0$), the current value of i mod 3, which is generated by a simple 3-state ring counter. For any specific value of i, we have a '1' on either i_0, i_1 or i_2, depending on whether i mod 3 is 0, 1 or 2. The inputs $s_1 s_0$ select the appropriate operand for the adder.

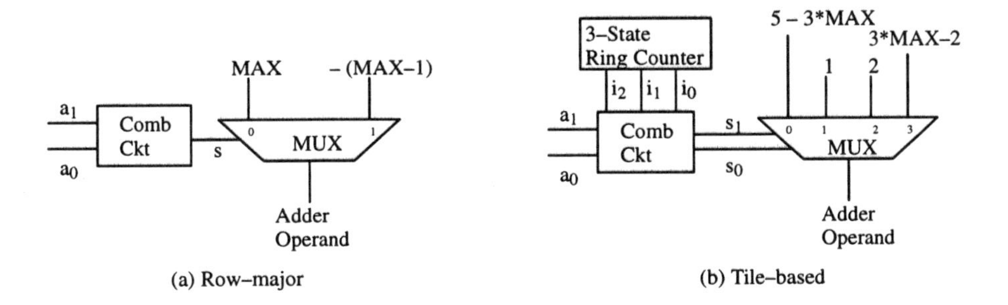

Figure 3.22. Implementation of Selector

Using the encoding 00, 01, 10 and 11 on controller output $a_1 a_0$ to indicate address A, B, C and D respectively being accessed, we arrive at the logic equations $s = a_1 \oplus a_0$, $s_1 = \overline{a}_1(a_2 \oplus i_0) + a_1(a_2 + i_2)$ and $s_0 = a_1 \oplus a_2 + a_1 i_2$.

Row-major and **Column-major** – Since the row-major scheme is the default mapping scheme used by most conventional synthesis tools, and column-major mapping is symmetrical, there is *no overhead* in the selector circuits resulting from these mapping schemes.

Tile-based – In Figure 3.21(c), the Selector circuit consists of a 2×1 MUX for row-major and a 4×1 MUX for tile-based. We compute the maximum difference in transition count per memory access, due to the 4×1 MUX as $4n$, where n is the bit-width (details in [PD95b]). [7] In addition, the Selector for tile-based mapping also has a 3-bit ring counter to keep track of i mod 3. The combinational circuits that generate the MUX control signals have a transistor count of 8 and 44 for row and tile mapping respectively – a difference of 36. Assuming 50% of the transistors change output [8] we get an equivalent transition count overhead of 18. The transition count in

[7] In brief, this is due to the fact that the *data inputs to the MUXes are constants*, and a bit-slice of the 4×1 MUX can only be as complex as an XOR gate ($6n$ transitions), whereas the 2×1 MUX can be as complex as an inverter ($2n$ transitions).

[8] This is usually an over estimation – note that the ring counter outputs never change during the inner loop execution

the ring counter can be ignored, since two bits change state only once every *outer* loop iteration.

Memory Address Decoder Power

To evaluate the effect of minimizing address bus transitions on the memory address decoder power, we conducted an experiment comparing the transition count in the address decoder for two sequences – one in which the address bus takes sequential values from 0 to $n-1$ (for various values of n) and another, in which the corresponding addresses are in Gray code sequence. Further details about the model of buffered address lines, split into row- and column- decoders etc. can be found in [PD95b]. The results show that the equivalent transition count in the decoder reduces by roughly 46%, i.e., a decrease in 50 % in the address bus transition count (Gray code vs. sequential)[9] led to a 46% decrease in internal transitions of the address decoder, demonstrating the close correlation between transition counts on the address bus, and the decoder circuitry.

3.2.1.4 Experiments. We conducted experiments for testing the efficacy of various memory mapping schemes on several examples taken from the Image Processing applications domain [PD95a]. We use the *cumulative count of the bit transitions* on the memory address bus during the execution of the algorithm as a power consumption metric for each memory mapping technique, and study the effects of the mapping schemes on transition counts by varying the dimension of the arrays.

Figure 3.23 shows the variation of the total transition count with respect to the dimension of the arrays for the SOR example. All arrays are of the form MAX $\times$ MAX, so MAX was varied along the x-axis from 50 to 1000 in steps of 10. The curves marked *Row-major* and *Row-major (Gray)* represent the total transition count for row-major mapping with and without the Gray code converter. Likewise for the *Column-major* and *Column-major (Gray)* curves. This illustrates the difference in transition counts due to the GCC alone. The curve marked *Tile-based* shows the transitions for the tile-based mapping.

For the SOR example, column-major mapping with the GCC does roughly as well as the tile-based mapping. In this case, the column-major mapping is still preferable, since it results in a simpler address generator compared to tile-based mapping, while performing almost equally well in terms of transition count.

Table 3.5 summarizes our experimental results by comparing the total transition counts for the row-major to the best mapping out of the ones we have considered, for seven examples from the Image Processing domain. The second and third columns

[9] Average bit-transition count for the sequential case is 2 [CLR92]. For Gray code, the average is 1 (by definition).

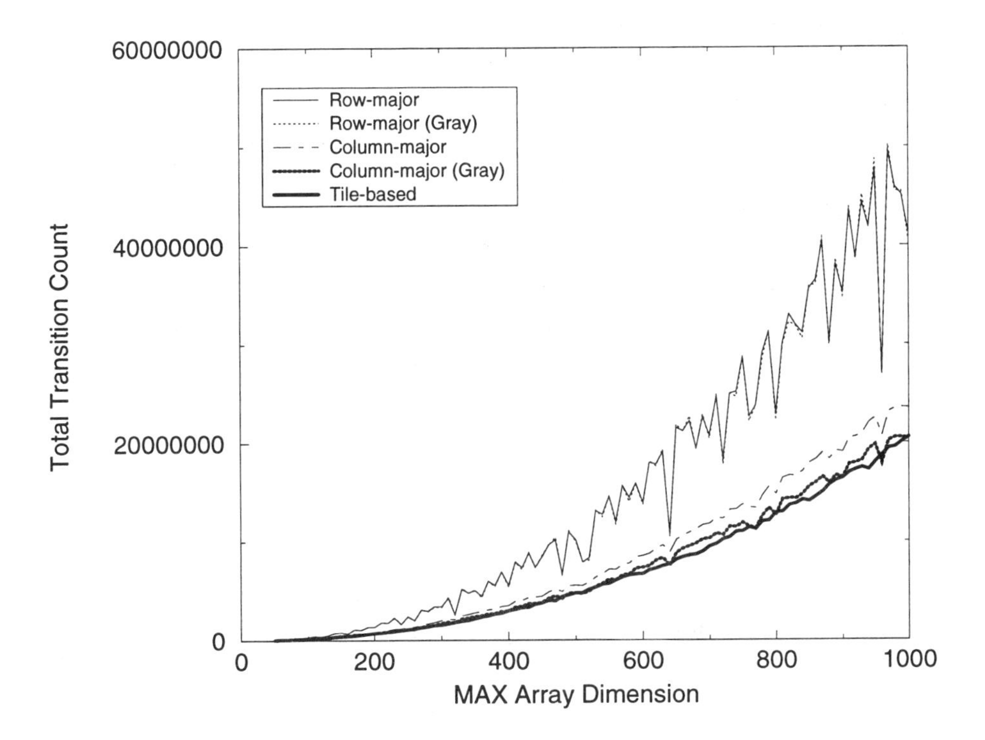

Figure 3.23. Transition Counts for the different mapping techniques for example SOR

Table 3.5. Transition Count Reduction for Image Processing Examples

Example	I/O Transition Count		% Red	Avg
	row-major	best	(1000)	% Red
Compress	11397910	6518370	42.8%	47.5%
GSR	28589669	28589669	0%	0%
Laplace	31739717	19653293	38.1%	41.3%
Lowpass	22413625	13218670	41.0%	41.2%
SOR	41153565	20438794	50.3%	51.9%

show the total I/O transition count for two mapping schemes – row-major and the best mapping predicted by Heuristic *SelectMapping*, with MAX = 1000. Column 4 shows the percentage reduction in I/O transition count for the specific case where MAX = 1000. Column 5 shows the average percentage reduction in transition count obtained by performing the same experiment for values of MAX ranging from 50 to 1000 in steps of 10. We observe a significant decrease in the memory address bus transition counts, indicating guaranteed power reduction for these examples.

Models of effective switched capacitances of the memory decoder, Gray code converter and Selector circuit (in address generator) were used to compute the reduction in transition count effected by tile-based mapping in comparison to row-major mapping, for the SOR example (with MAX=1000). Using average values of 20 fF and 10 pF for *on-chip* and *off-chip* capacitances [PD96, Bak88], the reduction in transition count (equivalent on-chip transitions) was determined to be 1772 for the ASIC *per memory access*, after considering the overhead introduced by the GCC and Selector. For the same execution, there was a 43% reduction in transition count in the memory decoder [PD95b].

3.2.2 Mapping Arrays into Multiple Memories

The technique presented in Section 3.2.1 can be extended to an architecture that consists of more than one single-ported memory modules of different sizes, as shown in Figure 3.24. We have a pair of data and address buses connecting the ASIC to each memory. This configuration is common in performance-critical systems where the required data access rate exceeds the maximum access rate possible when only a single memory can be accessed at any instant of time (i.e., multiple memories connected to a single data bus).

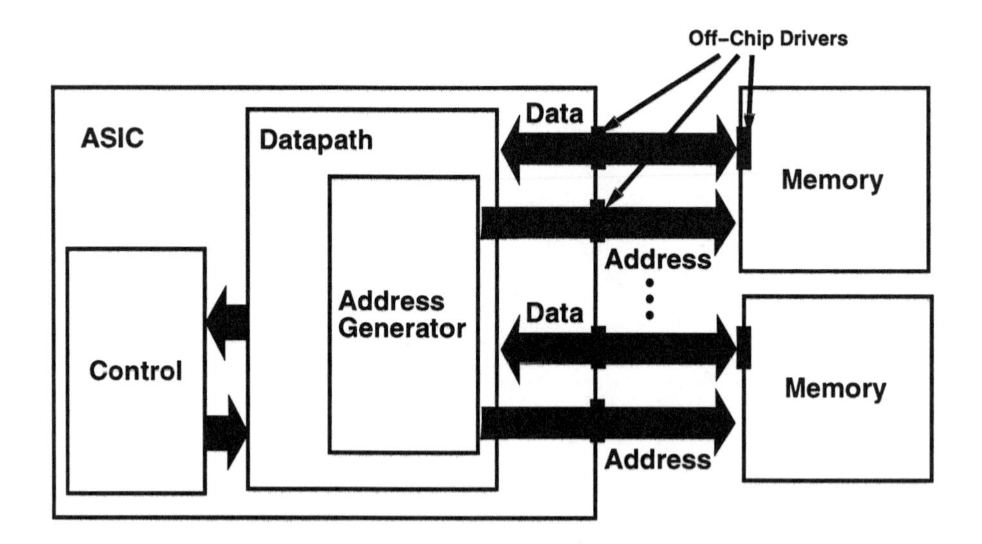

Figure 3.24. Architecture involving multiple memory modules

The approach is summarized in Figure 3.25. We first analyze the access patterns in the specification to determine the optimum partitioning of the arrays into multiple *logical array partitions*. For example, in Figure 3.25, array A_1 is split into logical array partitions K_1 and K_2. We then regroup the logical array partitions into *logical memories* based on criteria such as the possibility of interleaving (page 66). In Figure 3.25, logical array partitions K_1 and K_3 are merged into logical memory L_1. Finally, we map the logical memories into the available *physical memories*. The criterion here is minimization of the transition count overhead arising from mapping multiple logical memories to the same physical memory module. In Figure 3.25, logical memories L_1 and L_3 are assigned to physical memory P_1, whereas L_2 is assigned to P_2.

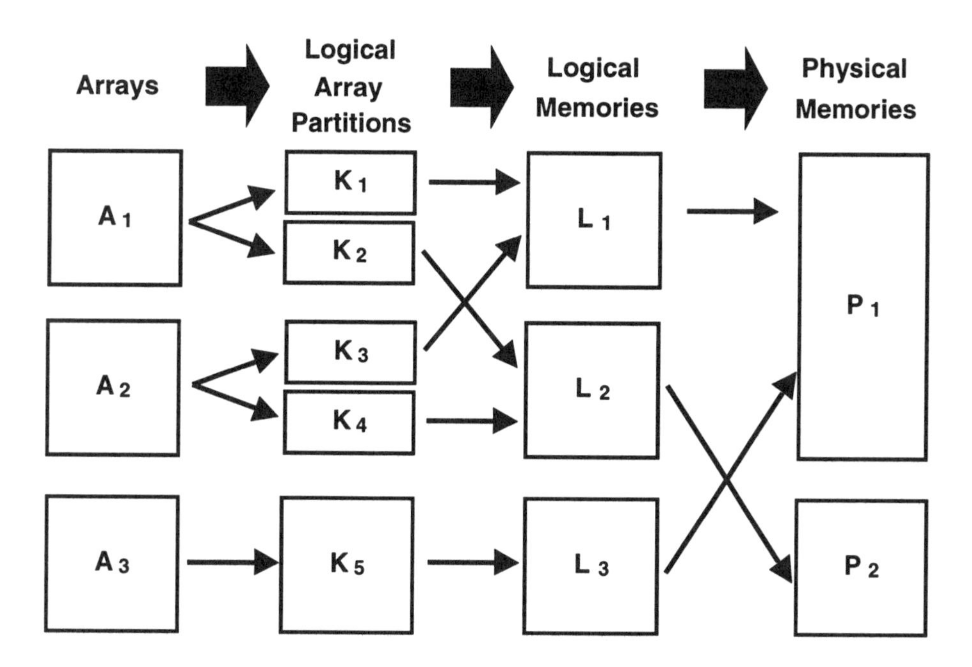

Figure 3.25. Mapping of Arrays to Multiple Physical Memories

3.2.2.1 Splitting into Logical Array Partitions. Regularity in behavioral access patterns, which is a common feature in most memory-intensive applications, especially those in the DSP/Image processing domain, allows us to extract the dimensions of the tiles (Section 3.2.1) for arrays accessed in loops. Figure 3.26 shows an array u, of dimensions 6×4, organized into four tiles. In the multiple memories scenario, for each array, we use the tile thus derived as the starting point and split it into as many logical array partitions as number of array elements in the tile.

Figure 3.26 illustrates the splitting of array u whose tile contour has already been established, into multiple array partitions. We have six array partitions corresponding to each element of the tile. There are as many elements in each array partition, as the number of tiles in array u. The rationale for this division is that if each array partition is mapped to a different logical memory, this ensures optimality in terms of bit-transitions on the address bus. For example, if $u[0][0]$ (in TILE1) is accessed in one loop iteration, we have $u[0][2]$, $u[3][0]$, and $u[3][2]$ being accessed in subsequent iterations (because the access pattern remains constant). The partitioning in Figure 3.26 results in all these four elements being mapped into the same logical partition, thereby ensuring that consecutive elements of the partition are accessed in each iteration. If the memory address is converted into Gray code, the address bus for each memory would have just one bit transitioning between consecutive iterations of the inner loop of the specification.

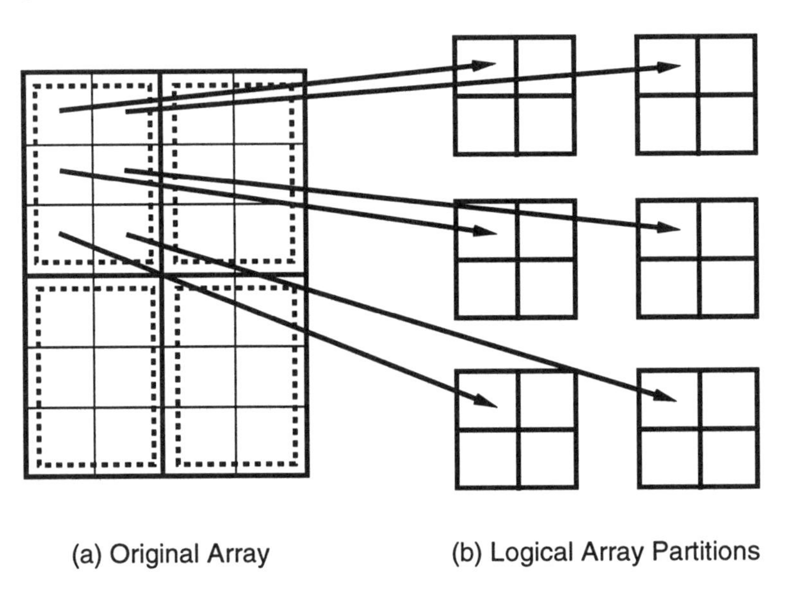

Figure 3.26. Partitioning of an array into multiple array partitions

3.2.2.2 Merging Logical Array Partitions. After partitioning the arrays in the specification to multiple logical array partitions, the next task is to group several array partitions into larger logical memories. This is because it might be prohibitively expensive to have a separate physical memory for each array partition. We consider the following properties while merging the array partitions:

Interleaving. It may be possible to interleave multiple arrays in the same physical memory without suffering any penalty in bit transitions.

Independence. If two arrays are not accessed in the same loop, they can be stored in the same logical memory.

Non-overlapping Lifetimes. If two arrays have non-overlapping lifetimes they can be stored in the same memory locations, since the same memory space can be reused for the two arrays, i.e., we need only one logical array partition for the two arrays.

Based on the above properties, we construct a *compatibility graph* G in which each vertex represents an array partition, and the presence of an edge indicates that the two partitions can be placed in the same physical memory with no transition count overhead (i.e., they are *compatible*). If two arrays are either independent, or can be interleaved, then we create an edge between the corresponding vertices.

After constructing the compatibility graph G, we apply a *clique partitioning* algorithm to divide the graphs into sub-graphs of array partitions. A clique is a fully connected subgraph of the original graph G. The significance of a clique is that all partitions in the clique can be placed in the same physical memory. An exact solution of the clique partitioning is known to be NP-complete [GJ79], so we employ an existing approximation algorithm [TS86] for this purpose. Each subgraph resulting from the clique partitioning corresponds to a logical memory.

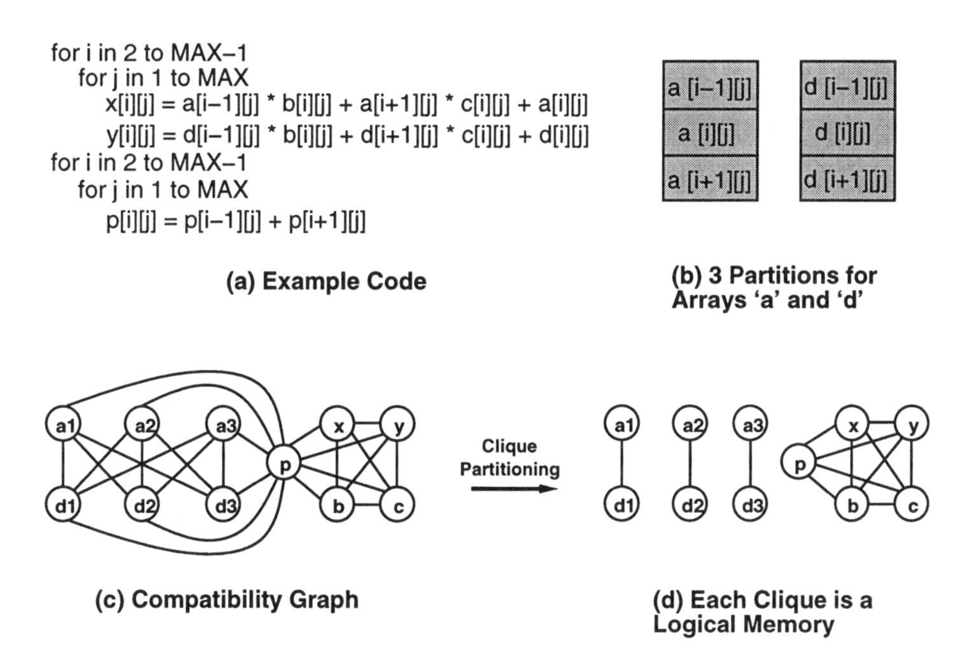

Figure 3.27. Deriving logical memories from behavior

Figure 3.27(a) shows an example behavior and Figure 3.27(c) shows the corresponding graph for the array partitions. The tile corresponding to arrays a and d have three elements each, since 3 elements of each array are accessed in the inner loop (Fig. 3.27(b)). This leads to three array partitions for the two arrays. The array p has edges to all other array partitions, since it is accessed in a different loop, thus the independence property allows it to be placed in the same logical memory with any of the other array partitions. One possible partitioning of this graph into cliques is shown in Figure 3.27(d). This leads to four logical memories, each consisting of clusters of array partitions.

3.2.2.3 Mapping into Physical Memory. The next step is to map the logical memories to available physical memory modules. There exists a trade-off between memory utilization and address bus transition count: mapping each logical memory into a separate physical memory might be prohibitively expensive, forcing multiple logical memories to be mapped into the same physical memory. However, this mapping leads to inefficiency in terms of transition count, because some components of two different cliques in the compatibility graph would be incompatible (otherwise they would be in the same larger clique). We introduce an additional user-supplied constraint, a *memory packing factor* f, which allows the user to control this trade-off. The constraint f, which represents the minimum fraction of each physical memory that needs to be filled, can, of course, be set to 1, indicating that all the physical memories should be full. On the other hand, if the value of f is relaxed to less than 1, then a better packing in terms of transition count might be achieved. If f is set to 0, then the resulting mapping is optimum with respect to transition count, but could be area-expensive, as memory space would be wasted.

The algorithm we use for mapping the logical memories into physical memories is a variant of the *Best-Fit Decreasing* heuristic for the *bin-packing* problem [CL91]. The general strategy is to consider the logical memories one at a time, largest first and assign it to a physical memory module based on its own size and the sizes of the available memory modules. We start with the largest logical memory (L), since this is a candidate that possibly accounts for large transition counts. If L is larger than all available physical memories, we implement it with (multiple copies of) the largest available physical memory module. If there is a remainder of lesser size, we add it to the list of logical memories and continue the mapping process. If there is at least one physical memory module of greater size than that of the logical memory under consideration, we map the logical memory into the largest physical memory module that satisfies the memory packing factor f. If the constraint f cannot be satisfied, the mapping is too expensive, and we need to compromise on transition counts by mapping more than one logical memory to the same physical memory. The details of the algorithm can be found in [PD95b].

3.2.2.4 Experiments. In our first experiment to determine the impact of multiple memories on transition count, we use a configuration of multiple memories in the examples, where there is one physical memory available for each logical memory. In other words, this is the best improvement possible over the single memory case, since the partitioning into logical memories represents the ideal mapping, according to our technique. We recall, from Section 3.2.2.3, that we try to avoid, as far as possible, mapping of multiple logical memories to the same physical memory.

Example	Transition Counts		Red.	Avg.
	Single Memory	Multiple Memories	(1000)	Red.
Compress	6518370	999002	6.5 X	6.6 X
GSR	28589669	8897227	3.2 X	2.9 X
Laplace	19653293	3997735	4.9 X	4.4 X
Lowpass	13218670	4993734	2.6 X	2.7 X
SOR	20438794	5997502	3.4 X	3.2 X

Table 3.6. Transition Count Reduction due to Multiple Memories

Table 3.6 shows a comparison of transition counts for five examples, all involving 2-dimensional arrays of dimensions (MAX $\times$ MAX), of the best mapping in the case of a single physical memory, and multiple physical memories. Columns 2 and 3 show transition counts for the different examples, with the value of MAX as 1000. Column 4 shows the number of times by which transition counts decreased as a result of using multiple memories. Column 5 shows the average reduction for all the different sizes of the arrays that were considered for each example (MAX was varied from 50 to 1000 in steps of 10). We observe that transition count could be reduced by a factor of between 2.7 and 6.6 times if multiple memories are considered.

Table 3.6 demonstrates the possible transition count reduction in the hypothetical case of physical memories of the same size as the logical memories being available. In practice, however, the designer may be constrained by a specific library of physical memory modules. In the second experiment, we use a specific library of memory modules of sizes 128 KBytes, 256 KBytes, 512 KBytes, 1 MByte and 2 MBytes, with the assumption that each array element in the arrays of the examples occupies one byte (data bus is 8 bits wide). Table 3.7 shows the transition counts obtained for the same five examples when the mapping is performed with the above library.

Table 3.7. Transition Count Reduction for a Specific Library

Example	Transition Counts (MAX = 1000)		Red.
	Simple Mapping	Our Mapping	
Compress	6701381	999002	6.7 X
GSR	19877346	11010165	1.8 X
Laplace	12884093	3997735	3.2 X
Lowpass	14060806	4993734	2.8 X
SOR	14118103	9145590	1.5 X

In Table 3.7, Column 2 shows the transition counts obtained for the case when MAX = 1000 using a simple algorithm that maps the arrays in the specification (in order of decreasing size) to the smallest memory that accommodates them. Column 3 shows the corresponding transition counts using our approach. Comparing Column 3 in Table 3.6 and Table 3.7, we note that for three of the examples (*Compress, Laplace* and *Lowpass*), the transition count in Experiment 2 was the same as the best case (one physical memory for each logical memory) transition count in Experiment 1 (i.e., our algorithm performed the optimal mapping). The transition count reduction factors shown in Column 4 indicate a significant reduction resulting from our approach, ranging from a factor of 1.5 to 6.7.

3.2.3 Mapping into Multiport Memories

The array mapping techniques discussed above can also be targeted at minimizing power consumption in architectures employing multiport memories. In this case, the goal is to assign arrays in a behavioral specification to memory locations, and ports of multiport memories, so as to minimize the transition count on the memory address buses when these arrays are accessed. We have a pair of data and address buses connecting the ASIC to each memory port. The datapath is connected to an m-port memory, with ports $P_0..P_{m-1}$, consisting of m data buses $D_0..D_{m-1}$ and m address buses $A_0..A_{m-1}$. The address generator generates the m addresses for the address buses $A_0..A_{m-1}$, so that up to m data words can be accessed in parallel.

The problem of mapping into a multiport memory requires the solution of two subproblems: (i) determine the mapping style to use for each array, and (ii) assign specific array elements accessed within behavioral loops to memory ports, such that the total transition count on all the address buses is minimized. The details are described in [PD97].

3.3 SUMMARY

In this chapter, we discussed optimization techniques for improving performance and power characteristics of synthesized designs for memory-intensive applications. The performance improvement is possible by using a refined model of memory access modes to pre-process the input specification. The power reduction is obtained through a judicious memory mapping strategy for behavioral arrays. Such techniques, which take the actual organization of modern memories into consideration, are necessary in order to incorporate memories in an automatic synthesis environment.

4 DATA ORGANIZATION: THE PROCESSOR CORE/CACHE INTERFACE

4.1 INTRODUCTION

Modern embedded systems increasingly rely on processors due to the advantages they offer in terms of flexibility, reduction in design time and full-custom layout quality [MG95]. The software part of the partitioned specification of Figure 1.2 is executed on these embedded processors, which are often available in the form of *cores*, to be instantiated as part of a larger system on a chip. This is feasible in current technology due to the relatively small area occupied by the processor cores, making the rest of the on-chip die area available for RAM, ROM, coprocessors, and other modules. Apart from the processors in the Digital Signal Processing domain (such as the TMS320 series from Texas Instruments), we also find microprocessors with relatively general purpose architectures available as embedded processors. An example of such a general purpose embedded processor is LSI Logic's CW4001 [ACG+95], which is based on the MIPS family of processors.

An embedded system's performance is greatly affected by interaction between the processor and external memory. General purpose embedded processors such as the CW4001 are equipped with on-chip instruction and data caches, which interface with larger off-chip memories. Since off-chip memory accesses usually stall the CPU execution for significant durations (each access could take 10-20 processor cycles, depending on the relative processor and memory access speeds), it is important to design the interface between cache and main memory carefully. Figure 4.1 highlights the components in the embedded system architecture that are addressed in this chapter.

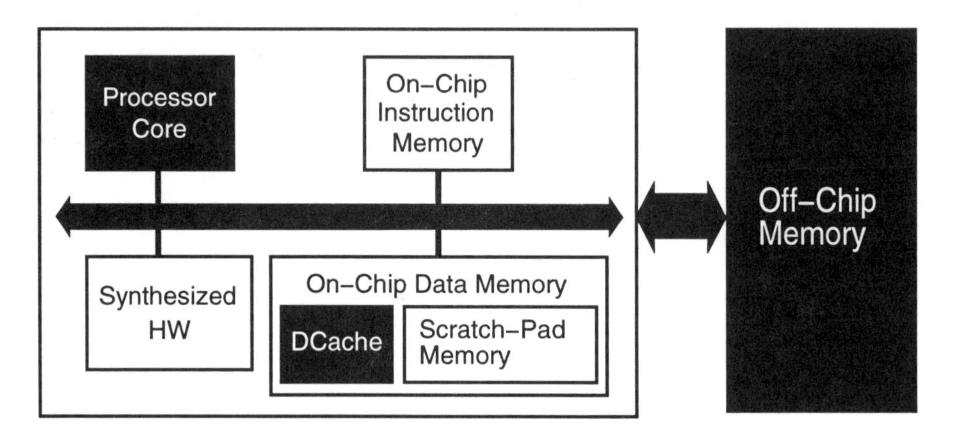

Figure 4.1. Shaded blocks represent the Processor Core/Data Cache/Off-chip Memory subsystem

Embedded system design is characterized by certain features that traditional compilers typically do not consider in their optimizations. For example, traditional compilers seldom take into account the specific cache parameters such as cache line size in their optimizations because fast compilation speed requirements preclude complex analysis procedures. However, in embedded systems, code generation can be tuned to the specific cache configuration to be used (or the specific configuration that is being currently explored) at the expense of longer compilation times. In this chapter, we study techniques that exploit this situation to organize data in memory in order to minimize data cache misses.

Code placement methods based on program traces for improvement of instruction cache performance have been reported [TY96a]. Rawat [Raw93] and Austin [Aus96] have addressed the problem of variable placement for improving cache performance. However, [Raw93] addresses only scalar variables, and [Aus96] treats scalar and array variables alike, thereby missing array placement opportunities based on their index expressions. We discuss a strategy for first placing scalar variables, and present an extension for analysis of array reference indices to determine the placement of arrays (Section 4.2). We then show how a data alignment strategy can be incorporated into *loop blocking*(Section 4.3).

4.2 MEMORY DATA ORGANIZATION

4.2.1 *Problem Description*

Consider a direct-mapped cache of size C ($C = 2^m$) words, with a cache line size M words, i.e., M consecutive words are fetched from memory on a cache read miss. In our formulation, we assume a *write-back* cache with a *fetch-on-miss* policy [Jou93], though the technique remains identical for other write policies, and is equally effective.

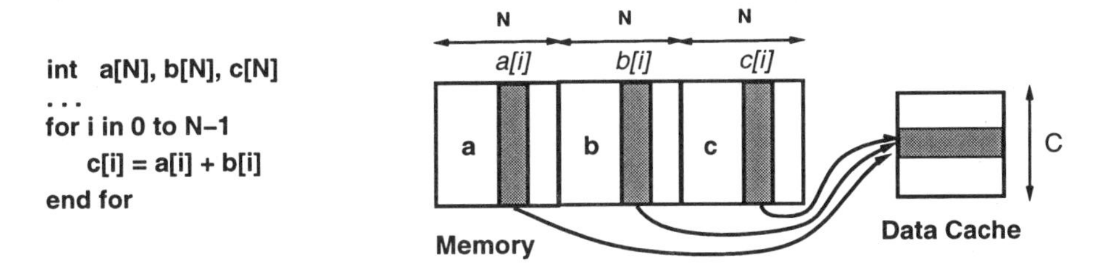

Figure 4.2. $a[i], b[i]$ and $c[i]$ map into the same cache line

Suppose the code fragment in Figure 4.2 is executed on a processor with the above cache configuration, where N is an exact power of 2, and $N > C$. Assuming one array element per memory word, let array a begin at memory location 0, b at N, and c at $2N$. Let $f(x)$ denote the cache line to which the program variable x is mapped. In a direct-mapped cache, the cache line containing a word located at memory address M, is given by: $(M \bmod C)/M$. In the above example, array element $a[i]$ is located at memory address: i. Similarly, we have $b[i] : N + i$ and $c[i] : 2N + i$. We find that the corresponding cache lines to which each of them will be mapped are:

$$
\begin{aligned}
f(a[i]) &= (i \bmod C)/M \\
f(b[i]) &= ((N + i) \bmod C)/M = (i \bmod C)/M (\textit{since } N \textit{ is divisible by } C) \\
f(c[i]) &= ((2N + i) \bmod C)/M = (i \bmod C)/M
\end{aligned}
$$

In other words, $a[i], b[i]$, and $c[i]$ are mapped onto the same cache line (Figure 4.2). The following sequence of events that take place in one loop iteration: a cache miss occurs while accessing $a[i]$; the line $f(a[i])$ is filled; accessing $b[i]$ now causes a miss, since, even if it was present in cache before, it was displaced by the last access to $a[i]$ (since $f(a[i]) = f(b[i])$); $f(a[i])$ is now filled by the line from b; the write to $c[i]$ also causes a miss, since $f(c[i]) = f(b[i])$; in a fetch-on-miss cache, this causes the same cache line to be displaced by elements of array c. The same cycle repeats in other iterations. In other words, every memory access results in a cache miss. Such memory access patterns are known to result in extremely inefficient cache utilization, especially because many applications deal with arrays whose dimensions are a perfect power of two [YJH+95]. In such situations, simply increasing the cache size does not present an efficient solution, because the cache misses are not caused due to lack of capacity. The conflict-misses can be avoided if the cache size C is made greater than $3N$, but this is often infeasible when N is large, and where feasible, there is an associated area and access time penalty incurred when cache size is increased; similarly, access time penalties are also increased if the cache associativity is increased. Reorganization of the data in memory results in a more elegant solution, while keeping the cache size relatively small.

One way of preventing the thrashing caused by excessive cache conflicts for this simple example is to pad M dummy memory words between two consecutive arrays that are accessed in an identical pattern in the loops. For this example, if array a begins at 0, array b begins at location: $N + M$ (instead of N) and array c begins at: $2N + 2M$ (instead of $2N$). We have:

$$
\begin{aligned}
f(a[i]) &= (i \bmod C)/M \\
f(b[i]) &= ((N + M + i) \bmod C)/M = (i \bmod C)/M + 1 \\
f(c[i]) &= ((2N + 2M + i) \bmod C)/M = (i \bmod C)/M + 2
\end{aligned}
$$

This ensures that $a[i]$, $b[i]$, and $c[i]$ are always mapped into different cache lines, and their accesses do not interfere with each other in the data cache. We extend this basic idea to organize scalars (Section 4.2.2) and arrays (Section 4.2.3).

4.2.2 Memory Organization of Scalar Variables

In considering the memory organization, we assume that the scheduling and register allocation of the code has already been performed, and the sequence of accesses to variables is fixed. The steps involved in organizing scalar variables into memory are, in brief:

1. Build a *Closeness Graph*, representing the degrees of desirability for keeping sets of variables in the same vicinity of main memory. For example, a set of M consecutive words in memory could be read into a cache line on a single cache miss.

2. Group the variables into clusters of M words, where M is the cache line size.

3. Build a *Cluster Interference Graph*, representing the non-desirability of mapping of clusters into the same cache line (to avoid conflict misses).

4. Assign memory locations to clusters.

4.2.2.1 Constructing the Closeness Graph. Let us first generate an *Access Sequence*, which is a graph representing memory references (loads and stores are treated alike) in the code. We create a node in the variable access sequence, corresponding to every memory access in the code, and a directed edge between nodes representing successive accesses. Figure 4.3(a) shows an example Access Sequence. The label 3 on edge $e \rightarrow a$ represents a loop with bound = 3. We then construct a *Closeness Graph* of the variables, which represents the degree of desirability for keeping sets of variables in the same vicinity in memory. For example, if M words accessed successively from memory are placed in consecutive locations, a single memory access could fetch them all into cache, thereby reducing up to $M - 1$ extra memory accesses caused due to *compulsory* misses.

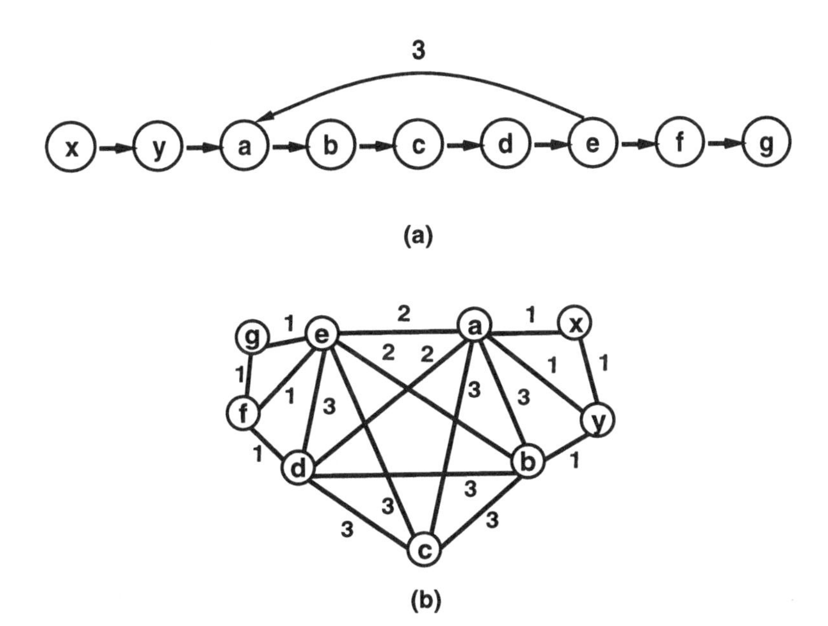

Figure 4.3. (a) Access Sequence (b) Closeness Graph for $M = 3$

We define the *distance* between two nodes u and v in the access sequence as: *distance*(u, v) = number of *distinct* variable nodes encountered on a path from u to v, or v to u (including u and v). The Closeness Graph *CG(V,E)* is constructed from the Access Sequence by first creating a node $v \in V$ for every variable in A, and initializing all edge weights $e(u, v) = 0$. For all occurrences u' of variable u during traversal of the Access Sequence, we examine a window of width M preceding and following u'. For all node instances v' (of variable v) in this window (where *distance*$(u', v') \leq M$), we update the edge weight $e(u, v) = e(u, v) + k$, where k is the number of times control is expected to flow between u' and v'. The required values of k in case of conditionals and loops could be obtained by using profiling information. We can also use the often-used simplifying assumption that branch probability is 0.5 for an *if*-statement, and that the loop bounds are always known at compile time. For a cache line size of M, the construction of the Closeness Graph requires, in the worst case, $O(Mn^2)$ time, where n is the number of nodes in the access sequence.

Figure 4.3(b) shows the Closeness Graph derived from the Access Sequence in Figure 4.3(a), with $M = 3$.

4.2.2.2 Grouping of Variables into Clusters. After constructing the Closeness Graph of the variables, the next step is to group the variables into clusters of M words, where M is the number of words in a data cache line. Intuitively, a higher edge weight in the Closeness Graph between two variables u and v represents a potential reduction in the number of memory accesses, if the two variables are stored close to each other, so that both can be accessed into the cache with a single memory read. In other words, it identifies an opportunity for avoiding compulsory misses in the data cache. To maximize the sharing of the cache lines by closely correlated variables, we now solve the following problem: *Partition the nodes of the Closeness Graph* CG *(i.e., the set of variables) into clusters of size* M, *so that the total weight of edges in all the clusters is maximized.*

An exact solution to the above problem has a computational complexity of $O(n^M)$, where n is the number of nodes in the Closeness Graph (i.e., no. of variables involved in memory accesses in the code), since there are $C^n_M (= O(n^M))$ ways of choosing the clusters. In most practical cache implementations, the line size M is usually less than 16. Hence, if the number of scalar variables is large, doing an exhaustive subset search of the Closeness Graph is usually impractical for $M \geq 4$. We employ the greedy heuristic *PerformClustering* in Figure 4.4 to perform the grouping of variables into clusters of size M. Essentially, the procedure attempts to identify clusters of M scalar variables whose memory accesses are temporally correlated, so that they can possibly be placed in consecutive memory locations. The heuristic uses the sum of incident edge weights as a criterion.

When procedure *PerformClustering* is applied on the graph in Figure 4.3(b), node b is selected first (line i). Next, line (ii) causes nodes c and d to be selected into the first cluster C_1. When we have equal T values for multiple nodes, we select one at random. Nodes b, c, d, and all connecting edges are now deleted. From the resulting graph, a, e, and g form the next cluster C_2. The final clustering is: $C_1 : [b, c, d]; C_2 : [a, e, g]; C_3 : [x, y, f]$.

To analyze the computational complexity of procedure *PerformClustering*, we note that the inner **while** loop eliminates one node every time it iterates, so it executes n

Procedure *PerformClustering*
Input: $CG(V, E)$ – Closeness Graph; M – Cache Line Size
Output: Set F – Set of clusters of size M
  for each vertex u in V
    Find the sum of incident edge weights $S(u) = \sum_{v \in V} e(u, v)$
  end for
  Let X = vertex set V and $Y = \phi$
  $--$ *Y keeps track of variables that are already assigned to clusters.*
  while $(X \neq \phi)$ **do**
    Let u = vertex $v \in X$ with maximum $S(v)$ (i)
    Create new cluster $C = \{u\}$
    while (size of cluster $C \neq M$) **and** $(X \neq \phi)$ **do**
      Let x be the variable $\in X$ with maximum value for T, (ii)
      where $T = \sum_{u \in C, v \in X - C} e(u, v)$ $--$ *i.e., x is the variable with maximum sum of edge weights with nodes already in C*
      $C = C \cup \{x\}$
      $X = X - \{x\}$
    end while
    Set $e(u, v) = 0 \forall (u \in C)$ or $(v \in C)$
    $--$ *i.e., delete all edges connecting to vertices in cluster C just formed*
    Update $S(v) \forall v \in X$
  end while
end Procedure

Figure 4.4. Procedure for clustering the variables of the Closeness Graph

times. In each iteration, it has to consider the edges from the $O(M)$ nodes in the cluster C, to the $O(n)$ remaining in X, i.e., each iteration takes $O(Mn)$ time. Similarly, the updating of $S(v)$ values requires $O(n)$ time for each vertex deleted. Since a total of n vertices are considered, the total complexity of the procedure is $O(Mn^2)$.

4.2.2.3 The Cluster Interference Graph. After grouping the variables into clusters of size M, we build a *Cluster Interference Graph (CIG)*, which represents the desirability to store clusters in memory, so that they do NOT map into the same cache line.

Each node in the Cluster Interference Graph represents one cluster of variables obtained in Section 4.2.2.2. A high edge weight between two nodes indicates a large number of conflict misses in the data cache, if the respective clusters were to map into the same cache line.

In procedure *BuildCIG* (Figure 4.5), we identify the clusters that should not map into the same cache line by assigning high edge weights to the edges between them.

Procedure *BuildCIG*
Input: A – Variable Access Sequence; F – Set of Clusters
Output: CIG – Cluster Interference Graph
  Convert the Variable Access Sequence A into a Cluster Access Sequence
    by renaming each node u in the sequence by the cluster C, where $u \in C$.
  Create a node in CIG for each cluster in F.
  Assign edge weight $e(u, v)$ between nodes u and v = the number of times
    the access to clusters u and v alternate along the execution path.
end Procedure

Figure 4.5. Procedure for generating the Cluster Interference Graph

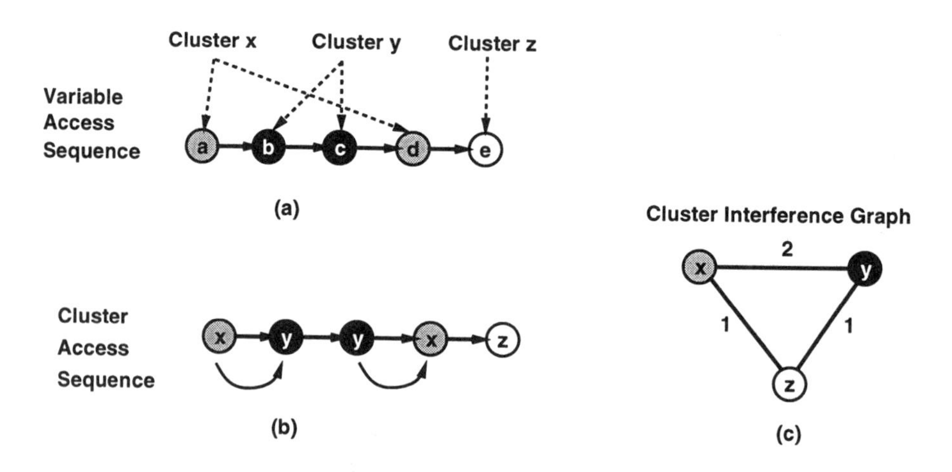

Figure 4.6. (a) Variable Access Sequence (b) Cluster Access Sequence (c) Cluster Interference Graph

Figure 4.6(b) shows the Cluster Access Sequence derived from the Variable Access Sequence shown in Figure 4.6(a), for $M = 2$. [1] In Figure 4.6(b), the pair of nodes x and y alternate twice in the execution path, due to the edges $x \rightarrow y$ and $y \rightarrow x$.

[1] Note that this clustering is for illustrative purposes only. If $a \rightarrow b \rightarrow c \rightarrow d \rightarrow e$ was the only execution chain in the program, a better clustering is $x : (a, b), y : (c, d), z : (e)$.

Hence, we have $e(x, y) = 2$ in the Cluster Interference Graph in Figure 4.6(c). The composition rules to be followed for conditionals and loops are identical to those used for building the Variable Access Sequence (Section 4.2.2.1).

For a Variable Access Sequence with n nodes, the process of generating a Cluster Access Sequence requires $O(n)$ time, since each node is renamed in constant time. A Variable Access Sequence with n nodes and a cache line size of M results in $r = \frac{n}{M}$ clusters. In order to obtain the Cluster Interference Graph from the Cluster Access Sequence, we maintain a 2-dimensional ($r \times r$) array, and update one row of the array (in $O(r)$ time) every time we encounter a node during the traversal of the Cluster Access Sequence. Hence, the total time required is $O(nr)$, i.e, $O(n^2/M)$ time.

4.2.2.4 Memory Location Assignment. The final assignment of variables to memory locations should take into account the clustering and conflict-penalty information in the Cluster Interference Graph. To minimize the conflict misses in the data cache during code execution, we need to ensure that cluster pairs with large edge weights do not map to the same cache line when we assign memory locations.

We define the *cost of a memory assignment* as follows:

$$MemAssignCost\,(CIG) = \sum_{x,y \in V(CIG)} e(x,y) \times P(x,y)$$

where $e(x, y)$ is the edge weight, and

$$P(x,y) = \begin{cases} 1 & \text{if memory locations for } x \text{ and } y \text{ map into the same cache line} \\ 0 & \text{otherwise} \end{cases}$$

Figure 4.7(b) shows a sample memory assignment for a CIG with six clusters (Figure 4.7(a)), on a cache with four lines. We note that cluster pairs (a, e) and (b, f) map into the same cache lines in a direct-mapped cache, where the cache line address is computed for memory location i, using the equation: $line = i \bmod 4$. The cost of the assignment (Figure 4.7(c)) is computed to be $1 + 3 = 4$, from the edge weights $e(a, e) = 1$ and $e(b, f) = 3$ for the two conflicting cluster pairs.

In order to minimize conflict misses, we need to solve the following *Cluster Assignment* problem:

> *Find an assignment of clusters in a CIG to memory locations, such that MemAssignCost (CIG) is minimized.*

The problem of assigning clusters in the Cluster Interference Graph to memory locations can be shown to be NP-hard, by showing a reduction from the Graph Colouring problem. The Graph Colouring problem can be stated as follows: is a given graph $G(V, E)$ k-colourable? That is, does there exist a colouring of vertices that uses $\leq k$ colours, in which no two adjacent vertices of G are assigned the same colour?

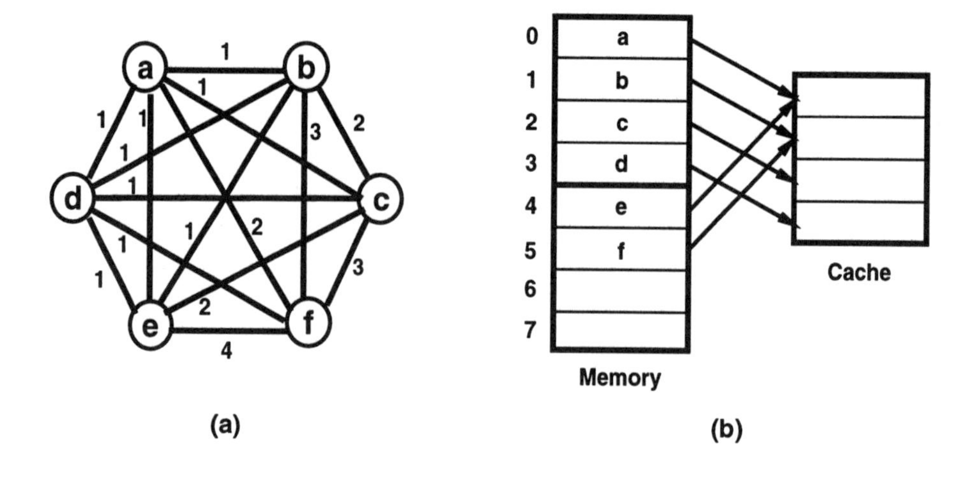

Conflicting Pairs	Cost
a, e	1
b, f	3

Total Cost = 1 + 3 = 4

(c)

Figure 4.7. (a) Cluster Interference Graph (b) Memory Assignment (c) Cost of Assignment

Theorem 1 *The Cluster Assignment Problem is NP-Hard*

Proof

Consider an instance of the Graph Colouring problem, where we are given a graph $G(V, E)$ where edge weights $e(u, v) = 1$ or 0, representing the presence or absence of an edge (u, v). The problem is to determine whether this graph is k-colourable. This problem is known to be NP-Complete [GJ79].

Now, if we had an optimal polynomial time algorithm A to solve the Cluster Assignment problem, we could interpret the same graph G as a Cluster Interference Graph – the input to algorithm A, where k is the number of cache lines. Let the cost of the optimal assignment computed by A for this graph be T. We can now arrive at the decision for k-colourability of G as follows:

If $T = 0$, then G is k-colourable, else G is not k-colourable. This follows from the equivalence between *two nodes being assigned the same colouring in G*, and *two clusters being assigned the same cache line in CIG*. Cost $T = 0$ for the memory assignment for CIG indicates an assignment, where no two nodes x and y such that $e(x, y) > 0$ are mapped into the same cache line. This is equivalent to no two adjacent nodes being assigned the same colour in the k-colourability problem for graph G.

In other words, if we had a polynomial-time algorithm to solve the Cluster Assignment problem optimally, we could also solve the Graph Colourability problem in polynomial time. Since the Graph Colourability problem is known to be NP-Complete, we have thus established the NP-hardness of the Cluster Assignment problem.

■

Procedure *AssignClusters* (Figure 4.8) is a greedy heuristic to solve the Cluster Assignment problem for a cache of size k that works along the same lines as the *PerformClustering* procedure.

We proceed to make the memory assignments *page* by *page*, where a page consists of k cache lines – the size of the data cache. Note that k consecutive clusters in memory will never conflict in cache, since, for the set of addresses $x = A, A+1, A+2, \ldots, A+k-1$, the cache mapping function $y = x \bmod k$ will always return distinct values. We define the cost of assigning cluster u to cache line i as $cost(u, i) = \sum_{v \in X} e(u, v)$, where X is the set of clusters that have already been assigned to cache line i. This cost is the sum of edge weights of u with all nodes that are already assigned to map into cache line i.

For the example CIG in Figure 4.7(a), the page size is $k = 4$ lines. When we apply procedure *AssignClusters* on this example, we first sort the vertices in decreasing order of the sum of their incident edge weights: $f(13), c(9), e(9), b(8), a(6)$ and $d(5)$. Clusters f, c, e, and b are placed into the first page P_0. While attempting to assign a into the second page P_1, we find: $cost(a, 0) = 2$ (since $e(a, f) = 2$), $cost(a, 1) = 1$, $cost(a, 2) = 1$, and $cost(a, 3) = 1$. Thus, we choose a line within page P_1 that minimizes the cost, and assign a to line 1 (the first available line with minimum *cost*). Cluster d has: $cost(d, i) = 1$ for all i, so we assign line 0 of P_1 to d. The final assignment is: P_0: $(0 - f; 1 - c; 2 - e; 3 - b)$ and P_1: $(0 - d; 1 - a)$.

To analyze the complexity of procedure *AssignClusters*, we note that the inner loop eliminates one node in each iteration, so it runs for a total of n iterations (where n is the number of clusters in the graph). Computation of the costs takes $O(n)$ time in this loop. Hence, the total complexity of the algorithm is $O(n^2)$. For an n-way associative cache, we use the same definition of *cost*, except that the cost remains zero for the first n clusters assigned to the same cache line.

```
Procedure AssignClusters
Input: CIG(V, E) – Cluster Interference Graph
Output: Assignment of Clusters to Memory Locations
    Sort the vertices of CIG in descending order of S(u)
    -- S(u) is the sum of edge weights incident on vertex u
    Let X be this sorted list of vertices
    while (X ≠ φ) do
        Create new page P in memory
        while (size of page P < k) and (X ≠ φ) do
            u = head of list X
            Assign u to line i of page P, where cost(u, i) is minimum
            over i = 0...k − 1
            Delete u from X
        end while
    end while
end Procedure
```

Figure 4.8. Procedure for assigning clusters to memory locations

At the end of this step, we have an assignment of all the scalar variables in the code to memory locations, that would minimize the compulsory and conflict misses in the data cache when they are accessed during program execution.

4.2.3 Memory Organization for Array Variables

In Section 4.2.2, we addressed the problem of memory location assignment for scalar variables, with the objective of minimizing compulsory and conflict misses in the data cache. In this section, we address the problem of organizing array variables into memory, with the intent of minimizing data cache conflict misses when the arrays are accessed at run time.

We solve the memory organization problem for arrays by first constructing an *Interference Graph* among arrays in the code, and then assigning memory addresses to each array by minimizing the possibility of cache conflicts with other arrays in the code. The problem of clustering of variables to avoid compulsory misses is not relevant in the case of arrays, as most arrays are usually much larger than a cache line – often larger than the cache itself.

4.2.3.1 Constructing the Interference Graph. In the case of arrays, we note that if two arrays A and B are accessed repeatedly within a loop, then there is a possibility that the accesses to A and B might cause conflict misses in the data cache (Section 4.2.1). The Interference Graph (IG) of arrays represents the possibility of cache conflicts between the arrays in the code.

We first create a node for each array in the specification. Next, we determine the arrays that are repeatedly accessed in each loop, and add the loop bound L to the edge weights between each pair of arrays. This signifies that a total of L cache conflicts could possibly arise between each pair of arrays during execution of this loop. The resulting IG gives us a criterion to prioritize the order in which we assign memory addresses to arrays. Procedure *BuildArrayIG* (Figure 4.9) outlines the construction of the Interference Graph.

Procedure *BuildArrayIG*
Input: Code with array accesses
Output: Interference Graph *IG* of arrays
    Create a node u for every array u in the code
    Initialize edge weights $e(u, v) = 0$ for all u, v
    for all (innermost) loops *l* in the code **do**
        Let L be the loop bound of loop *l*
        Let X = set of all arrays accessed in *l*
        Update $e(u, v) = e(u, v) + L$ for all $u, v \in X$
    end for
end Procedure

Figure 4.9. Procedure for building Interference Graph for arrays

The complexity of procedure *BuildArrayIG* is $O(Ln^2)$, where L is the number of loops, and n is the number of arrays in the code.

In Figure 4.10(b), we show the Interference Graph derived from the code shown in Figure 4.10(a). The first loop causes $e(a, b) = 8$. Subsequently, the second loop adds 16 to $e(a, b)$, $e(a, c)$, and $e(b, c)$.

The IG helps identify the order in which the memory address assignment to arrays should be done.

4.2.3.2 Memory Assignment to Array Variables. In solving the problem of memory assignment of array variables, we assume that the loop bounds and array dimensions are known at compile time. We also assume that a unidimensional array of N elements is stored in N consecutive memory locations, and multidimensional arrays are stored in row-major format. (The issue of selection of a good storage technique for multi-dimensional arrays is addressed in Chapter 3 and in [CL95]). The memory assignment

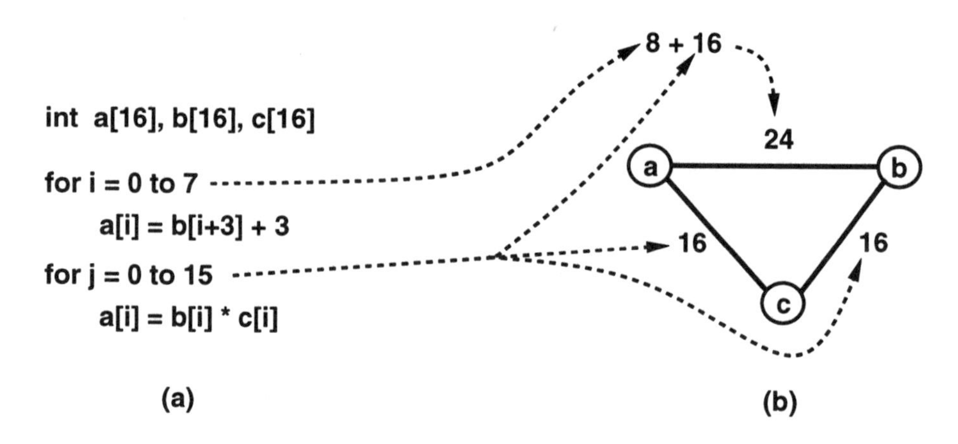

Figure 4.10. (a) Code showing arrays accessed in loops (b) Interference Graph: edge weights contributed by loop bounds

problem is NP-hard, because the degenerate case, when the array dimension = 1, itself happens to be NP-hard (Section 4.2.2.4).

From the Interference Graph, we use the $S(u)$ values for each node u (defined in Section 4.2.2.2) to determine the order of assignment of arrays. $S(u)$ signifies the relative importance of the nodes, because a higher $S(u)$ indicates that u could possibly be involved in many cache conflicts.

Central to the technique we use for memory assignment of arrays, is a computation of the *cost* of assigning an array (u) to begin at a specific memory address A. This cost is equal to the *expected number of cache conflicts with all arrays that have already been assigned*, if u were to begin at A. Note that if the first element of u is fixed at address A, all the other elements of u are automatically assigned their respective locations.

To determine whether two specific array accesses in the same loop will map into the same cache line (i.e, cause cache conflict miss), we perform a symbolic evaluation of the equality checking function. Two memory locations X and Y will map into the same cache line in a direct-mapped cache with k lines (M words per line), if the following condition holds:

$$\left(\left\lfloor \frac{X}{M} \right\rfloor - \left\lfloor \frac{Y}{M} \right\rfloor\right) \bmod k = 0$$

i.e., $(\lfloor X/M \rfloor - \lfloor Y/M \rfloor)$ is an integral multiple of k, which resolves to:

$$(nk - 1) < \frac{X - Y}{M} < (nk + 1) \tag{4.1}$$

where n is any integer.

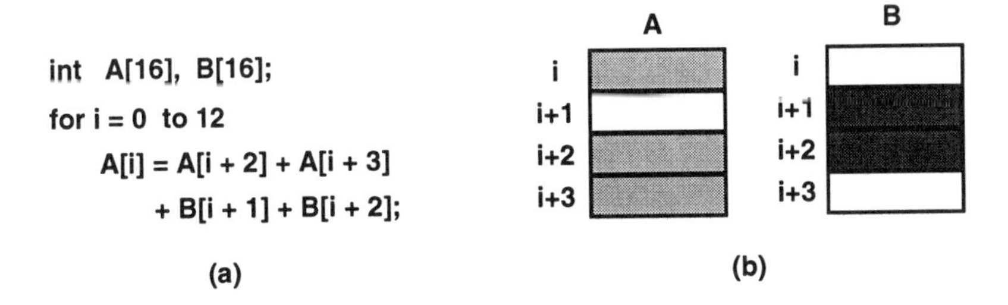

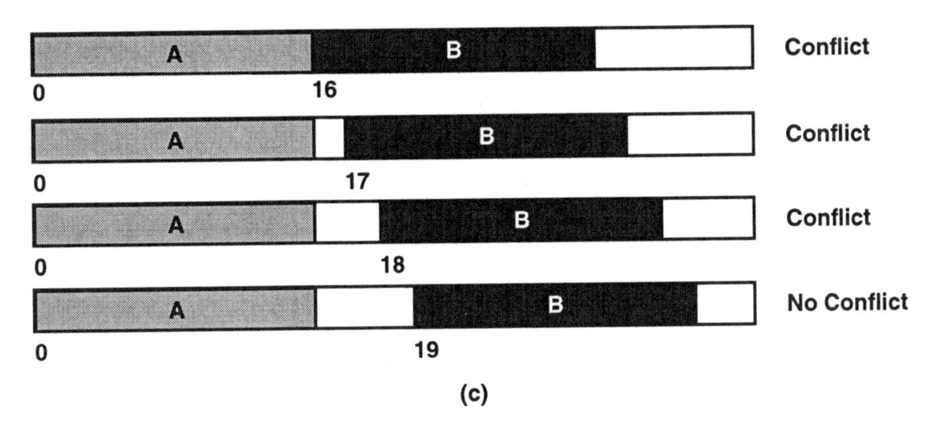

Figure 4.11. (a) Sample Code (b) Accessed Array Elements (c) Conflicts during assignment to B

Figure 4.11(a) shows a sample code, for which the array access pattern (in the i-th iteration) is shown in Figure 4.11(b). In the general case, for an array W starting at location T, $W[i]$ is located at address: $T + i$. Let us attempt to arrive at a memory assignment for array B, by first assuming that A begins at address 0, and the parameters for the cache are: $M = 4, k = 4$.

We first attempt to place B at address 16 (since A and B are arrays of 16 words each, and A begins at 0). Specifically, let us determine whether $B[i+1]$ and $A[i+3]$ map into the same cache line. $A[i+3]$ is located at address: $(i+3)$ and $B[i+1]$ is located at address: $16+i+1 = 17+i$. To check whether Condition (4.1) is satisfied, we note that: $(X - Y)/M = ((17+i) - (i+3))/4 = 14/4 = 3.5$, which satisfies Condition 4.1 for $n = 1$, as we have: $4 - 1 < 3.5 < 4 + 1$. Thus, $A[i+3]$ and $B[i+1]$ would conflict in cache. Similarly, we notice conflicts for assigning addresses 17 and 18 to B. However, address 19 does not entail any cache line conflicts between elements of A and B accessed in the same iteration (Figure 4.11(c)). Thus, we assign address 19 to B. The three locations 16, 17, and 18, could either be left blank, or occupied by scalar variables.

Clearly, the symbolically evaluated expression: $(X - Y)/M$, might not always reduce to a constant, because X and Y could be arbitrary functions of any variable in the code. If the expression does not resolve to a constant, then we conclude that the two arrays do not conflict.

We now formalize the strategy we presented in the example above into a procedure to perform the memory assignment of arrays. We first describe the cost function *AssignmentCost* (Figure 4.12) that returns the expected number of conflicts when an array is tentatively assigned a specific location. In practice, the inner **for**-loop is terminated if the assignment cost is zero for any assignment.

Function *AssignmentCost*
Input: u – array under test; A – proposed start address; Access Sequence;
  Array assignments already completed; *IG* – Interference Graph
Returns: Expected number of cache conflicts for this assignment
  Initialize *cost* = 0
  for all $v_l | e(v_l, u) \neq 0$, v_l already assigned
  –– *i.e., all assigned arrays that have an edge with u in IG*
    for each loop (bound L) in which accesses to v_l and u occur
      w = no. of times control alternates between elements of v_l and u
      that map into the same cache line, using Condition (4.1)   (i)
      cost = *cost* + $w \times L$ –– $w = 0$ *if there is no conflict*
    end for
  end for
  return *cost*
end Function

Figure 4.12. Cost function for expected number of conflicts

In line (i) of *AssignmentCost*, the number of times control alternates between v_l and u is determined by Condition (4.1). w represents the number of cache misses in the loop due to conflict between v_l and u – this is determined from the Access Sequence. Procedure *AssignArrayAddresses* (Figure 4.13) outlines the strategy for determining the addresses for each array.

Function *AssignmentCost* requires $O(P)$ time, where P is the total number of distinct array accesses in the code, since, in the worst case, $O(P)$ comparisons might be required. The worst case complexity of procedure *AssignArrayAddresses* could be $O(nkP)$, where n is the number of nodes (arrays), k is the number of cache

Procedure *AssignArrayAddresses*
Input: *IG* – Interference Graph, k – no. of cache lines
Output: Assignment of addresses to all arrays (nodes in *IG*)
Address $A = 0$
Sort nodes in *IG* in decreasing order of $S(u)$ (sum of incident edge weights)
Let the list of nodes be: $v_0 \ldots v_{n-1}$
for $i = 0 \ldots n-1$
Initialize cost $c = \infty$
min $= 0$ $\quad$ -- *keeps track of cache line with minimum mapping cost*
for $j = 0 \ldots k-1$
if *AssignmentCost*$(v_i, A + j) < c$ **then**
c = *AssignmentCost*$(v_i, A + j)$
min $= j$
end if
end for
Assign address $(A + min)$ to first element of v_i
$A = A + min +$ *size*(v_i) $\quad$ -- *updating A for next iteration*
end for
end Procedure

Figure 4.13. Procedure for assigning addresses to arrays

lines. However, in real behaviors, we have observed that the assignment cost in loop $j = 0 \ldots k-1$ tends to converge to zero very soon (typically less than 2 or 3 iterations), because the number of different array elements accessed in inner loops of code is usually small.

This completes the memory address assignment of scalar and array variables in the behavior.

4.2.4 Mapping into Set-Associative Caches

Sections 4.2.2 and 4.2.3 considered the layout of data in memory in order to minimize misses in a direct-mapped cache.We now consider the case of mapping of data into a *set-associative* cache. In an a-way associative cache with total capacity C elements and cache line size L, the number of lines $n = C/L$, and number of sets $s = n/a$. A word residing in block number M in memory would reside in set number: $(M \bmod s)$ in the set-associative cache, as opposed to line number: $(M \bmod n)$ in a direct-mapped cache of the same capacity.

Clearly, the clustering technique (Section 4.2.2) to minimize compulsory misses remains identical for the associative cache, since the clustering depends only on the cache line size. However, after the Cluster Interference Graph is obtained, the memory assignment phase differs from that used in the direct-mapped cache.

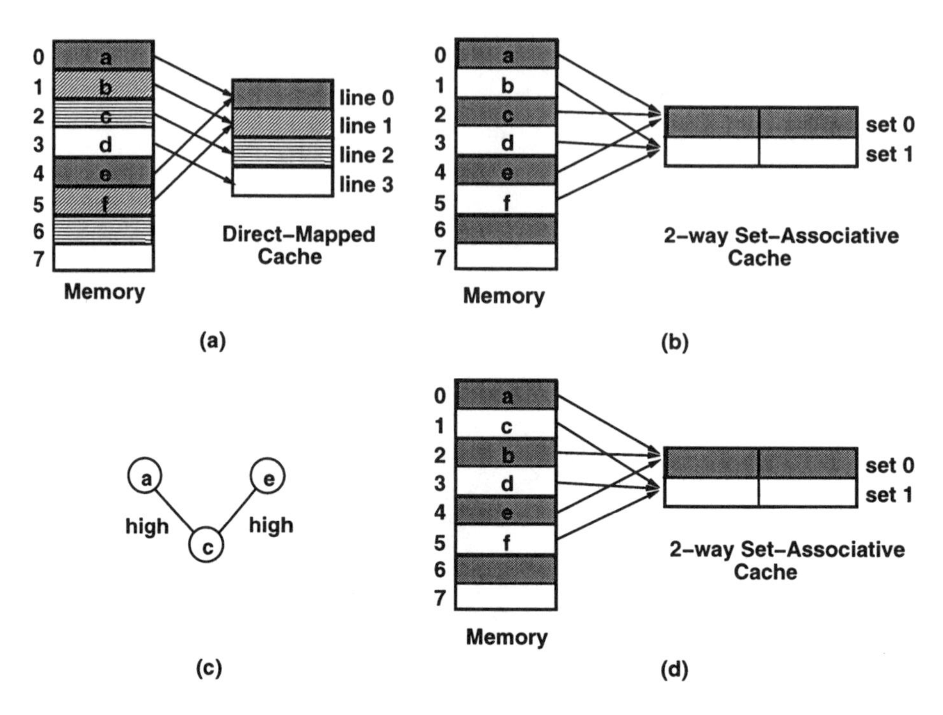

Figure 4.14. (a) Mapping into 4-element direct-mapped cache (b) Mapping into 4-element 2-way set-associative cache (c) Unfavourable Interference Graph for 2-way set-associative cache (d) Optimal data placement for 2-way set-associative cache

We show that the optimal memory assignment for direct-mapped caches is not necessarily optimal for set-associative caches, with an example in Figure 4.14. Figure 4.14(a) shows the distribution of six elements: a, b, c, d, e, and f located consecutively in memory into locations of a direct-mapped cache with size 4 elements, and cache line size = 1 element. Let us assume that this memory assignment is optimal for the given direct-mapped cache, as determined from Section 4.2.2.4. Figure 4.14(b) shows the distribution of the same memory elements in a 2-way associative cache of size 4 elements. We notice that elements a, c, and e map into the same set in the cache. This mapping could lead to undesirable conflicts, for example, when the interference graph is of the form shown in Figure 4.14(c). The weights of edges $e(a, c)$ and $e(c, e)$ are high, indicating expensive conflicts if the pairs compete for a cache location.Figure 4.14(c) shows only the relevant portion of the interference graph.

Other edges (with low edge weights) are not shown. The conflicts are avoided in the direct-mapped cache (Figure 4.14(a)), because c is mapped into a different line than a and e.

To obtain an optimal memory assignment for a set-associative cache with s-sets, we consider the page size $= s$ while doing the memory assignment in Section 4.2.2.4 (instead of page size $= n$ in a direct-mapped cache). This decision is based on the observation that any s consecutive clusters in memory will always be mapped into distinct sets in the cache, and hence will not conflict. This argument is analogous to that used for direct-mapped caches – any n consecutive clusters in memory will always map into distinct cache lines. In the example shown in Figure 4.14, we have $n = 2$ and $s = 2$. To obtain the optimal memory assignment for the set-associative cache, we compute the memory assignment with page size $= 2$. Figure 4.14(d) shows an example solution. We observe that in this assignment, c is mapped into a different set from a and e, thereby avoiding the expensive conflicts indicated earlier.

In summary, the data layout strategy for an a-way set-associative cache is identical to that for a direct-mapped cache of the same capacity (n), except that in the memory assignment phase, we use a page size of $s = n/a$, instead of n for the direct-mapped cache.

4.2.5 Experiments

In this section, we describe experiments on several benchmark examples to validate our memory data organization strategy. For our experiments the CW4001 simulator kit from LSI Logic was employed, executing on a SUNSparc5 workstation, that simulates code and reports performance statistics of code generated for execution on LSI Logic's CW4001 embedded processor. The specific performance statistics of interest for the experiments are the *data cache hit ratio* and the *total number of processor cycles* needed for execution of the code on the embedded processor.

We first give a brief description of the benchmark examples for the experiments, followed by the results on a direct-mapped and a 2-way set-associative data cache of the same capacity. For these experiments, the *read* and *write latencies* (the number of processor cycles required to access an external memory word) are assumed to be 5 processor cycles. These values are generally among the lowest possible (i.e., for the fastest memories). Since the performance difference widens even more for higher penalties (higher access times), the observed improvements are the minimum possible (i.e., most conservative). A Write Buffer [PH94] of size 4 words was used in the data cache, which was configured to be a *write-through* cache employing *no-fetch-on-write, no-write-allocate* write miss policy, with a cache line of size 4 words. We finally present a study on the variation in performance for different memory latencies.

4.2.5.1 Benchmark Examples. All the examples on which we consider are benchmark code kernels used in image processing, telecommunication, and other applications in the DSP and scientific domain. Table 4.1 presents a profile of these benchmark examples. Column 1 shows the benchmark names; Column 2 gives the number of scalar variables in the code; Column 3 gives the number of array variables; and Column 4 gives the number of array references in the code for each example. For the examples where the array sizes were not specified, we used a size of 64 for one-dimensional arrays, and 16×16 for two-dimensional arrays. *SOR* and *Laplace* [PD95a] are C-implementations of algorithms frequently used in DSP applications such as image processing. *Dequant, leaf_plus* and *leaf_comp* are modules from the MPEG decoder application. *Idct* is the Inverse Discrete Cosine Transform routine, used in many DSP applications including the MPEG. *FFT* is the Fast Fourier Transform routine, also popular in the DSP domain. *Matrix_add, inner_product* and *tri_diagonal_elim* are frequently used in routines involving two dimensional arrays treated as matrices. *Hydro, eqn_of_state, 1D_PIC, implicit_cond, 2D_hydro, ordinates_transport*, and *2D_implicit_hydro* are other code kernels of typical scientific applications, all of which form part of the Livermore Loops set of benchmark suite.

Table 4.1. Profile of Benchmark Examples: Data Organization

Benchmark	No. of Scalar variables	No. of Array variables	No. of array accesses in code
SOR	4	7	13
Laplace	2	2	10
dequant	7	5	5
FFT	20	4	20
idct	20	3	9
leaf_comp	5	3	3
matrix_add	2	3	3
hydro	6	2	6
inner_prod	2	3	2
tri_diag_elim	2	3	4
eqn_of_state	5	4	10
1D_PIC	4	12	33
implicit_cond	11	7	10
2D_hydro	8	9	44
ordinates_transport	7	9	15
2D_implicit_hydro	5	6	11

4.2.5.2 Experiment on Direct-Mapped Cache. Our first experiment is on the CW4001 processor core configured with a 1 KByte instruction cache and 256 Byte direct-mapped data cache with a 16-byte cache line. Figure 4.15 presents a comparison of the data cache hit ratios for the *Unoptimized* and *Optimized* memory data organizations, i.e., for the straightforward organization, and our technique respectively. In all the examples (except *Laplace*), we notice that the difference in the hit ratios is substantial. On an average, we observe a difference of 48 % in the data cache hit ratios.

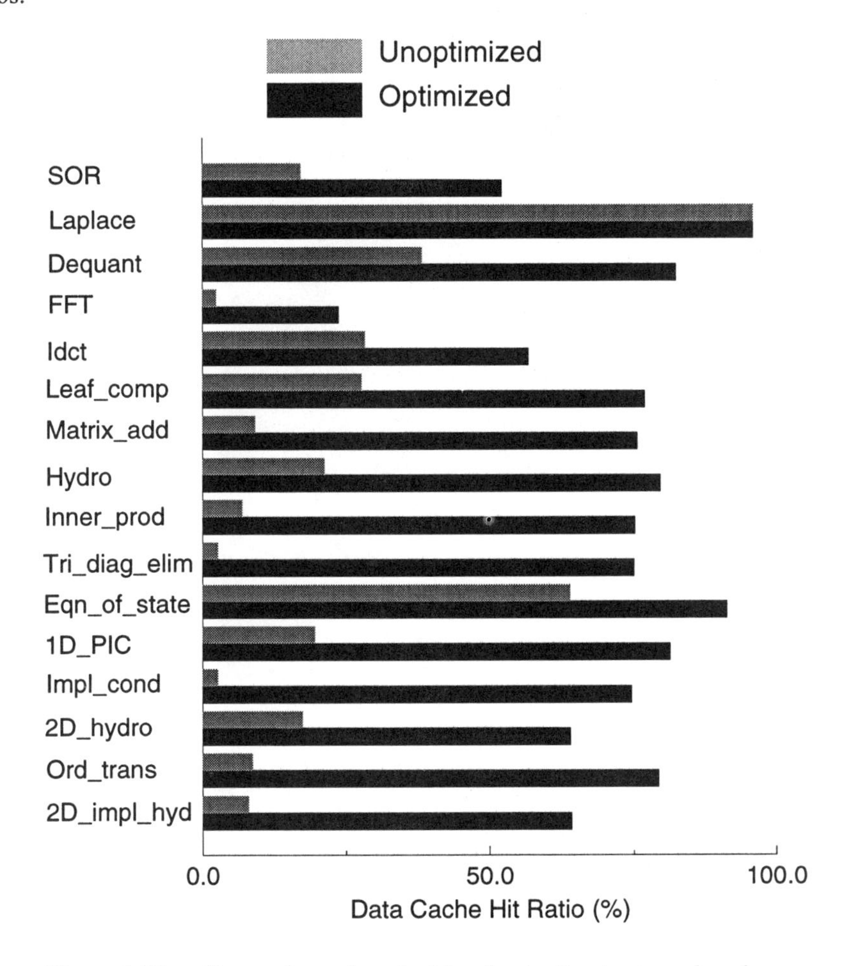

Figure 4.15. Comparison of cache hit ratios in direct-mapped cache

Figure 4.16 shows the speed-up in execution time on the CW4001 embedded processor resulting from our data layout technique, which is defined as:

$$\text{Speedup} = \frac{\text{\# Processor cycles for } \textit{Unoptimized} \text{ case}}{\text{\# Processor cycles for } \textit{Optimized} \text{ case}}$$

The *Laplace* example, for which no speed-up is observed, involves only two arrays, *image1* and *image2*, with *image1* being read and *image2* written. Since the cache

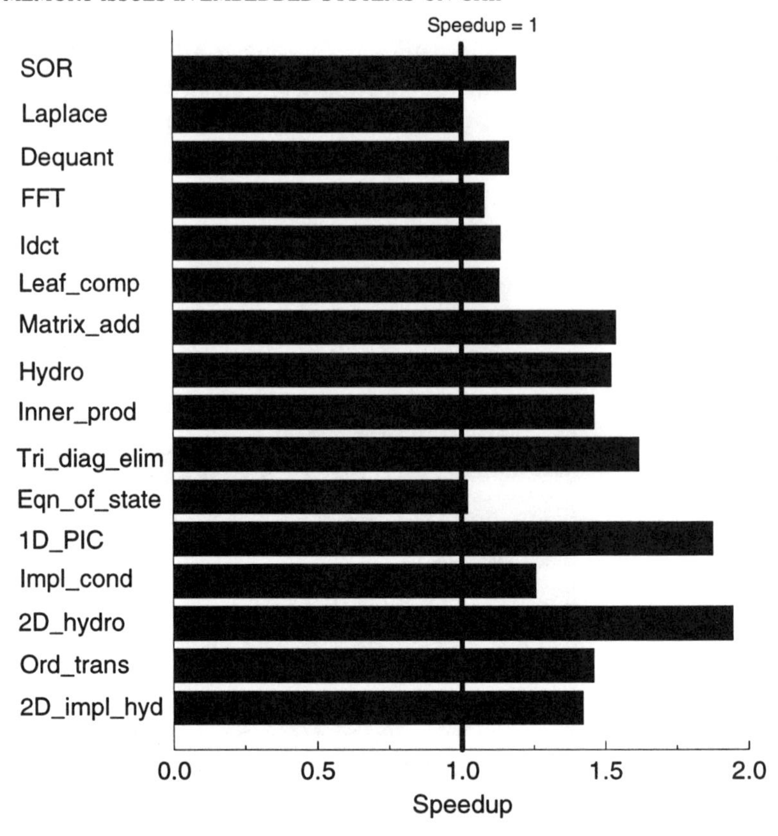

Figure 4.16. Comparison of speed-up in direct-mapped cache

implements a *write-through, no-fetch-on-write* policy, the two arrays do not conflict in the cache. Hence, the cache performance is good in the unoptimized case and data organization does not lead to any improvement. In all the other examples, cache conflicts are reduced by our data layout technique, leading to reductions in execution time.

There is a significant reduction in the total cycle time for most of the applications. The average reduction in the total cycle time over all the examples was 34%. The reason why the percentage difference tends to be lower than the difference between data cache hit ratios is that, the total execution time also factors in all the other determinants of program execution speed, most notably, the instruction cache. The instruction cache has an identical performance in both the optimized and unoptimized cases, since in this work, we have concentrated on the performance of the data cache alone. Further, the speed-up is a function of the memory latency; it increases with memory latency (Section 4.2.5.4).

Table 4.2 gives a summary of the overhead incurred by our data organization strategy, in terms of memory space and compilation time. Columns 2 and 3 show the memory space required for storing the variables in the *Unoptimized* and *Optimized* cases respectively. Column 4 shows that in all cases, the wasted memory space due to the data layout is negligible in comparison to the size of the data. On an average, the overhead incurred is 2.17%. Columns 5 and 6 show the compilation times required for the *Unoptimized* and *Optimized* strategies. As we observe in Column 7, the compilation time overhead for performing the data placement analysis was a maximum of 7.4%. On an average, the compilation time overhead was only 4.0%.

Table 4.2. Memory and Compilation time Overheads

	Memory Required (Bytes)			Compilation Time (seconds)		
Benchmark	*Unopt*	*Opt*	Overhead	*Unopt*	*Opt*	Overhead
SOR	7184	7408	3.1%	0.51	0.54	5.1 %
Laplace	2056	2072	0.8%	0.42	0.44	4.1 %
dequant	2332	2364	1.4%	0.40	0.42	4.0 %
FFT	2128	2176	2.3%	0.58	0.59	1.9 %
idct	1616	1632	1.0%	0.50	0.52	3.4 %
leaf_comp	1556	1588	2.1%	0.41	0.43	3.4 %
matrix_add	3080	3112	1.0%	0.38	0.40	4.9 %
hydro	1048	1080	3.1%	0.35	0.36	2.6 %
inner_prod	1036	1052	1.6%	0.39	0.40	3.1 %
tri_diag_elim	1544	1576	2.1%	0.44	0.45	3.2 %
eqn_of_state	2068	2116	2.3%	0.42	0.44	4.2 %
1D_PIC	3628	3724	2.6%	0.51	0.54	5.6 %
implicit_cond	3088	3264	5.7%	0.45	0.47	3.1 %
2D_hydro	8096	8224	1.6%	0.92	0.95	3.6 %
ordinates_transport	4636	4764	2.8%	0.53	0.56	4.1 %
2D_implicit_hydro	5396	5476	1.5%	0.44	0.48	7.4 %

Most of the performance improvement reported in Figure 4.15 arises from the placement of arrays. This is because the number of scalar variables present in the examples was relatively low. The miss ratio reductions as a result of our data placement strategy are significantly higher than those observed in [Aus96]. This is due to several reasons. Firstly, we have a scheme to analyze array indices and determine the relative placement of arrays in memory. Secondly, most of the benchmark examples are code kernels exhibiting serious cache conflicts in the absence of data alignment. Thirdly, we have used a relatively smaller cache, thereby increasing the possibility of conflicts. The conflicts in the *Unoptimized* case will decrease when larger data caches are used. However, the experiments show that good cache performance can be obtained even for smaller caches by data alignment. Finally, the array sizes in the code kernels of the benchmarks were multiples of the cache size. It is possible that for other randomly chosen array sizes, the *Unoptimized* case would incur fewer conflicts. However, the special case of array sizes being close to a multiple of the cache size occurs frequently – e.g., image processing applications frequently involve images of size 512×512 or 1024×1024, which are very likely to result in multiples of cache sizes, since cache sizes are typically a power of 2. The data organization technique helps prevent pathological cache conflicts irrespective of array sizes.

4.2.5.3 Experiment on 2-way Set-Associative Cache. Our second experiment was performed on the CW4001 processor configured with a 1 KByte instruction cache and 256 Byte, 2-way set-associative data cache with a 16-byte cache line. Figures 4.17 and 4.18 summarize the data cache hit ratios and execution time speed-up respectively on the new cache configuration.

A comparison of Figure 4.15 and Figure 4.17 shows that the cache hit ratios are generally higher in the set-associative cache, for both *Unoptimized* and *Optimized* cases. However, our data layout technique still results in a significant (average of 29%) improvement in hit ratios for the set-associative cache. Our technique improves upon the straightforward strategy (*Unoptimized*) in all the examples, with the exceptions of *Laplace* (where the performance remains identical) and *Idct* (where there is a 0.6% degradation). The *Laplace* example is discussed in Section 4.2.5.2. In the case of *Idct*, only two arrays are read in the inner loop, which results in the 2-way set-associative cache performing well in the *Unoptimized* case due to the lack of any conflicts.

Figure 4.18 shows an average speed-up in execution time of 21% for our data layout technique over the *Unoptimized* technique. The memory space wasted for the set of examples was also low, and averaged 2.17%.

4.2.5.4 Variation of Performance with Memory Latency. In our experiments in Sections 4.2.5.2 and 4.2.5.3, we assumed a memory latency of 5 processor cycles for fetching a cache line from external memory. This latency is generally among the minimum possible in modern memories. The widening disparity between processor and memory speeds indicates that this value will continue to rise in the future. The execution time increases for higher latencies, resulting in higher speed-ups for our technique.

In Figure 4.19, we show the variation of execution time with memory latency for the *SOR* benchmark example. The graph shows the variation of execution times for the *Unoptimized* and *Optimized* cases, for memory latency values of all integral multiples of the processor cycles between 5 and 25. We notice that the speed-up increases from 22% for memory latency = 5 cycles, to 33% for memory latency = 25 cycles.

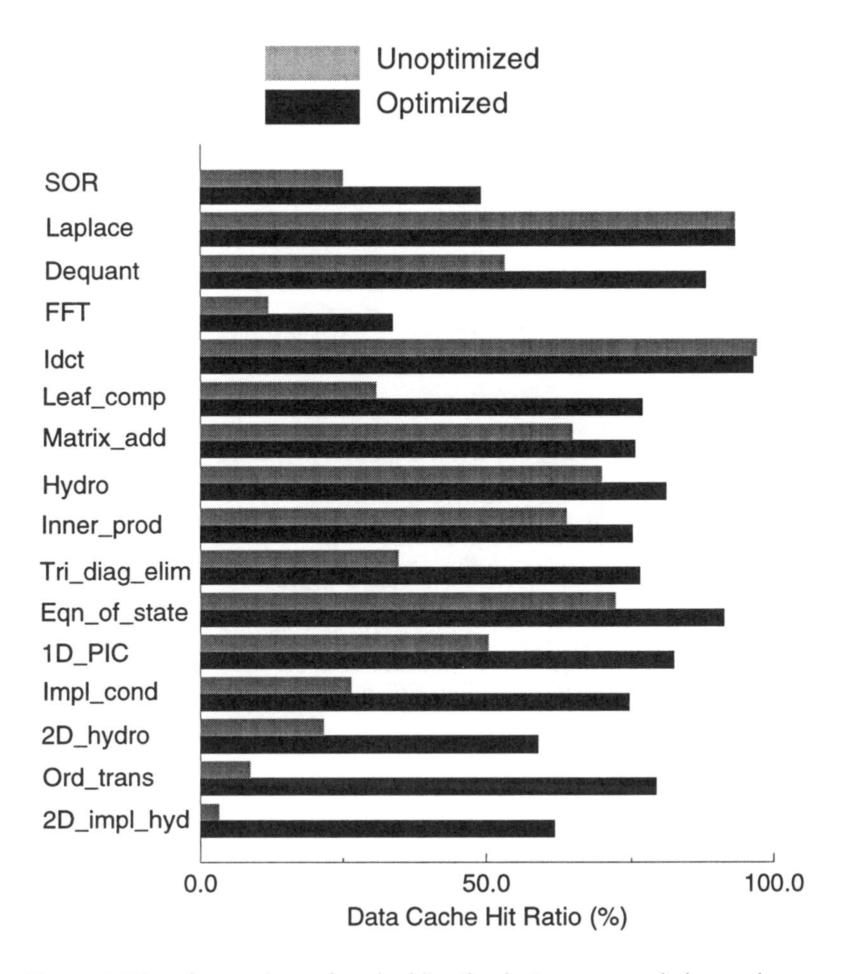

Figure 4.17. Comparison of cache hit ratios in 2-way associative cache

In summary, the execution time speed-up due to our data layout strategy increases with increasing memory latency.

4.3 DATA ALIGNMENT FOR LOOP BLOCKING

We now present a data alignment technique DAT[2] that judiciously uses padding, which when combined with tile size selection for loop blocking, results in improved, and more stable cache behavior.

[2] This section is the result of joint work with Professor Hiroshi Nakamura of the University of Tokyo

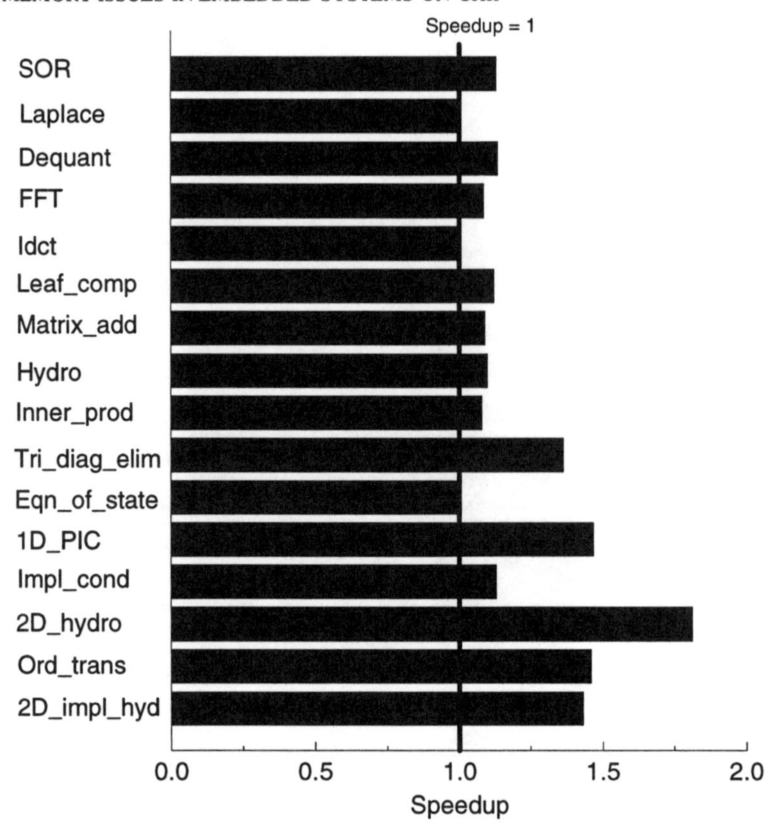

Figure 4.18. Comparison of speed-up in 2-way associative cache

4.3.1 Loop Blocking

Reduction of cache capacity misses, which occur when reusable data is evicted from the cache due to cache size limitations, is achieved by *loop blocking* (or *loop tiling*) [Wol89] – a well known compiler optimization that helps improve cache performance by dividing the loop iteration space into smaller *blocks* (or *tiles*). This also results in a logical division of arrays into tiles. The *working set* for a tile is the set of all elements accessed during the computation involving the tile. Reuse of array elements within each tile is maximized by ensuring that the working set for the tile fits into the data cache. Conflict misses, which occur when more than one data element competes for the same cache location, are a consequence of limited-associativity caches. Cache conflicts in the context of tiled loops are categorized into two classes:

Self-interference: cache conflicts among array elements in the same tile; and

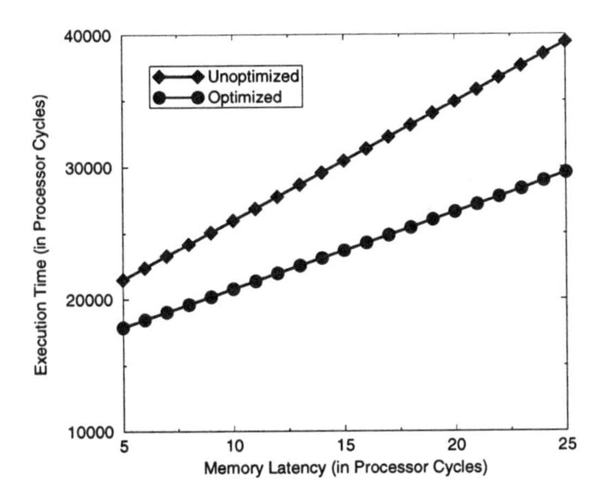

Figure 4.19. Variation of execution time with memory latency for SOR example

Cross-interference: cache conflicts among elements of different tiles (of the same array) or different arrays. Conflict misses can seriously degrade program performance; their reduction is generally addressed during the selection of tile sizes.

An important problem in the context of blocking is to determine the shape and size of the tile. Several techniques for determining the tile size have been studied in the past[LRW91, Ess93, CM95].

Lam, et al. [LRW91] reported the first work modeling interferences in direct-mapped caches with a study of the cache performance of a matrix multiplication program for different tile sizes. They present an algorithm (LRW) for computing the tile size, which selects the largest *square* tile that does not incur self-interference conflicts. This strategy is effective for reducing cache misses, but is somewhat sensitive to the array size – the tile sizes vary widely with small changes in array sizes. For instance, in the matrix multiplication example, for a 1024-element cache, a 200×200 array results in a 24×24 tile, whereas a 205×205 array results in a 5×5 tile. The 5×5 tile makes inefficient utilization of the 1024-element cache and degrades performance due to loop overheads introduced by tiling.

Esseghir [Ess93] presents a tile size selection algorithm (ESS) for one-dimensional tiling, which selects a tile with as many rows of the array, as would fit into the data cache. This algorithm cannot exploit the benefits of two-dimensional tiling, and also does not consider cross-interference among arrays in its computation of tile sizes.

Coleman and McKinley [CM95] present a technique (TSS) based on the Euclidean G.C.D. computation algorithm for selecting a tile size that attempts to maximize the cache utilization while eliminating self-interferences within the tile. They incorporate the effects of the cache line size, as well as cross-interference between arrays. The tile sizes generated by TSS are, like LRW, also very sensitive to the array dimension. For the matrix multiplication example on a 1024-element cache, a 200×200 array results in a 41×24 tile, whereas, a 205×205 array results in a 4×203 tile. Since the working set in both cases is large, the cache utilization of TSS is good. However, if the working set gets too close to the cache size, cross-interferences, which are handled with a probabilistic estimate in TSS, begin to degrade the performance. We discuss this issue in Section 4.3.4.

The data alignment technique proposed by Lebeck and Wood [LW94] involves padding each record (in an array of records) to the nearest *cache line boundary* to prevent accesses to individual records from fetching an extra cache line from memory, thereby decreasing the number of cache misses. Further, when two or more arrays accessed in a loop have a combined size smaller than the cache, they are laid out in memory so as to map into different regions of the cache. However, their technique does not address programs with tiled loops.

Padding is a data alignment technique that involves the insertion of dummy elements in a data structure for improving cache performance. In this section, we discuss DAT, a data alignment technique based on padding of arrays, that ensures a relatively stable, and in most cases, better cache performance, in addition to maximizing cache utilization, eliminating self-interference, and minimizing cross-interference. Further, while all previous efforts are targetted at programs characterized by the reuse of a single array, we also address the issue of minimizing conflict misses when several tiled arrays are involved. Experiments on several benchmarks with varying memory access patterns demonstrate the effectiveness of our technique and its stability with respect to changing problem sizes.

4.3.2 Motivating Example

We illustrate the effect of the padding data alignment technique on the cache performance of the Fast Fourier Transform (FFT) algorithm. Figure 4.20(a) shows the core loop of FFT [EK91], highlighting the accesses to array *sigreal*.

Figure 4.21(a) illustrates the distribution of the accesses in the *sigreal* array, for two iterations of the outer loop (indexed by l), with array size $n = 2048$ words, and an example 512-word direct-mapped cache with 4 words per line. In the first iteration of the outer loop ($l = 0$), we have accesses to the set of pairs (*sigreal*$[i]$, *sigreal*$[i + 1024]$), all of which conflict in the cache, because, in general, the pairs (*sigreal*$[i]$, *sigreal*$[i + 512k]$) will map to the same cache location for all integral k. Similarly, we observe similar severe conflicts between the pairs (*sigreal*$[i]$, *sigreal*$[i + 512]$) in the second iteration ($l = 1$).

```
Algorithm FFT_Original
Input: n, wreal[n * 2^(n-1)],
    wimag[n * 2^(n-1)]
Inout: sigreal[2^n], sigimag[2^n]
le = 2^n; windex = 1; wptrind = 0;
for (l = 0; l < n; l++) {
    le = le/2;
    for (j=0; j < le; j++) {
        wpr = wreal [wptrind];
        wpi = wimag [wptrind];
        for (i=j; i < 2^n; i += 2*le) {
            xureal = [sigreal [i]];
            xuimag = sigimag [i];
            xlreal = [sigreal [i + le]];
            xlimag = sigimag [i + le];
            [sigreal [i]] = xureal + xlreal;
            sigimag [i] = xuimag + xlimag;
            tr = xureal - xlreal;
            ti = xuimag - xlimag;
            [sigreal [i + le]] = tr * wpr -
            ti * wpi;
            sigimag [i + le] = tr * wpi -
            ti * wpr;
        } /* loop i */
        wptrind += windex;
    } /* loop j */
    windex = windex * 2;
} /* loop l */
```

(a)

```
Algorithm FFT_Padded
Input: n, wimag[n * 2^(n-1)],
    CACHE_SZ, LINE_SZ
Constant: NO_LINES =
    CACHE_SZ / LINE_SZ
Inout: sigreal[2^n + 2^n/NO_LINES],
    sigimag[2^n + 2^n/NO_LINES]
le = 2^n; windex = 1; wptrind = 0;
for (l = 0; l < n; l++) {
    le = le/2;
    le' = le + le / NO_LINES;
    for (j=0; j < le; j++) {
        wpr = wreal [wptrind];
        wpi = wimag [wptrind];
        for (i=j; i < 2^n; i += 2*le) {
            i' = i + i / NO_LINES;
            xureal = [sigreal [i']];
            xuimag = sigimag [i'];
            xlreal = [sigreal [i' + le']];
            xlimag = sigimag [i' + le'];
            [sigreal [i']] = xureal + xlreal;
            sigimag [i'] = xuimag + xlimag;
            tr = xureal - xlreal;
            ti = xuimag - xlimag;
            [sigreal [i' + le']] = tr * wpr -
            ti * wpi;
            sigimag [i' + le'] = tr * wpi -
            ti * wpr;
        } /* loop i */
        wptrind += windex;
    } /* loop j */
    windex = windex * 2;
} /* loop l */
```

(b)

Figure 4.20. (a) Original FFT algorithm (c) Modified FFT algorithm

The pathological conflicts above can be prevented by the judicious padding of the *sigreal* array with dummy elements. Figure 4.21(b) shows a modification of the storage of the *sigreal* array, with 4 dummy words (size of one cache line) inserted after every 512 words. This ensures that $sigreal[i]$ and $sigreal[i + 2^n]$ never map into the same cache line. We note in Figure 4.21(b) that for $l = 0$ and 1, $sigreal[i]$ and $sigreal[i + le]$ map into consecutive cache lines. For $l \geq 2$, they map into different

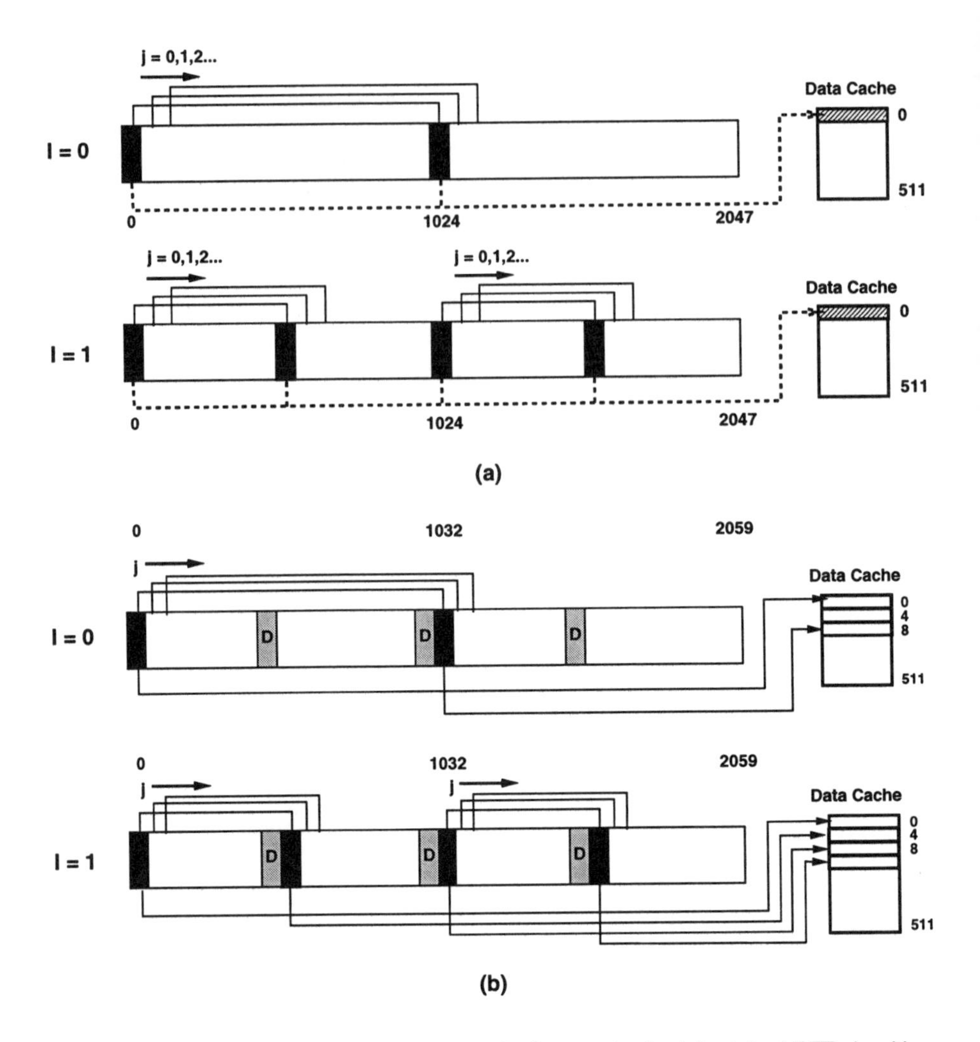

Figure 4.21. Distribution of array accesses in data cache for (a) original FFT algorithm (b) modified FFT algorithm

regions of the cache (as before), but the padding now avoids the conflicts mentioned before. [3] In order to implement the padding strategy, the array index computation would now change minimally, resulting in the code shown in Figure 4.20(b). In the computation of i' and le' in the *FFT_Padded* algorithm, NO_LINES (= CACHE_SZ / LINE_SZ) is usually a power of two, and consequently, the respective integer divisions result in shift operations.

[3] The *sigimag* array, which has an identical access pattern to *sigreal*, is prevented from conflicting with *sigreal* by adjusting the distance between the two arrays, i.e., by inserting padding between them.

A comparison of the execution times of the original FFT algorithm in Figure 4.20(a) with that of Figure 4.20(b) shows a speedup of 15% on the SunSparc-5 machine. This shows that the time spent in the extra computation involved in calculating the new array indices (i' and le') is small in comparison to the time saved by preventing the cache misses.

Note that the cache conflict problem identified in the FFT algorithm of Figure 4.20(a) could be solved by using a 2-way set-associative cache instead of a direct-mapped one. However, we also need to avoid a similar conflict in array *sigimag* (and between arrays *sigreal* and *sigimag*), so a 4-way set-associative cache would be required to avoid conflicts in this example. In general, cache access time considerations will not allow the use of a cache with arbitrarily large associativity. The padding technique helps avoid conflicts in a direct-mapped cache, and consequently, also avoids conflicts in any associative cache of the same size. We discuss the the possible effects of the padding data alignment technique on the rest of the program in Section 4.3.5.

4.3.3 Data Alignment Strategy

We now describe DAT, a technique for data alignment of two-dimensional arrays. In Section 4.3.3.1, we describe the procedure for selecting the tile sizes. For the selected tile sizes, we compute the required padding of arrays for avoiding self-interference within the tile in Section 4.3.3.2. In Section 4.3.3.3, we present a generalization of the technique to handle the case when a tiled loop involves several arrays with closely-related access patterns.

4.3.3.1 Tile Size Computation. The tile size selection procedure can be summarized as follows: select the largest tile for which the working set fits into the cache. Note that the working set here could include some elements outside the chosen tile. For computing the working set size, we use the formulation used in [CM95]. The objective of this strategy is to maximize the cache utilization, but it would normally lead to self-interference among elements of the chosen tile. To prevent this self-interference, we adjust the arrays with an appropriate padding (Section 4.3.3.2).

The de-coupling of the two steps (tile size selection and padding) makes the technique both efficient and flexible. The efficiency is derived from the fact that the cache is never under-utilized. The flexibility arises from the ability to eliminate self-interferences independent of the tile size. The important consequence of this flexibility is that the tile size can now be optimized independently, without regard to self-interferences. For instance, if one-dimensional tiling were considered the best

Procedure *SelectTile*
Input:
TypeX, TypeY, ShapeX, ShapeY,
Cache Size (C), Associativity (A), Cache Line Size (L)
Output:
SizeX, SizeY /* No. of rows and columns of selected tile */

Case 1 (*ShapeX* = *ShapeY* = 0):
/* Common case: no application-specific info. available */
$SizeX = SizeY = \mathbf{max}\{k | (k = L \cdot t, t \in I) \wedge (ws(k, k) \leq C)\}$
Case 2 (*TypeX* = CONSTANT **and** *TypeY* = VARIABLE):
/* Fixed number of rows */
$SizeX = ShapeX$
$SizeY = \mathbf{max}\{k | (k = L \cdot t, t \in I) \wedge (ws(SizeX, k) \leq C)\}$
Case 3 (*TypeY* = CONSTANT and *TypeX* = VARIABLE):
/* Fixed number of columns */
$SizeY = ShapeY$
$SizeX = \mathbf{max}\{k | (k \in I) \wedge (ws(k, SizeY) \leq C)\}$
Case 4 (*TypeX* = CONSTANT and *TypeY* = CONSTANT):
/* Tile size already determined */
$SizeX = ShapeX; SizeY = ShapeY$
Case 5 (*TypeX* = VARIABLE and *TypeY* = VARIABLE):
/* *ShapeX*/*ShapeY* gives ratio of #rows and #cols in tile */
$(SizeX, SizeY) = (ShapeX \cdot t, ShapeY \cdot t)$ such that
$t = \mathbf{max}\{k | (k \in I) \wedge (ws(ShapeX \cdot k, ShapeY \cdot k) \leq C)\}$
end Procedure

Figure 4.22. Procedure for selecting tile size

choice for an application, it could be easily incorporated into our technique. This cannot be handled in TSS and LRW.[4] We discuss the de-coupling further in Section 4.3.5.

Figure 4.22 describes procedure *SelectTile* for computing the tile sizes. In addition to the common case where we select the square tile, we also allow the user to specify the *shape* of the tile in cases where additional information about the application is available. To achieve this, the procedure takes the parameters: *TypeX, TypeY, ShapeX,* and *ShapeY*. In the common case (case 1: *ShapeX* = *ShapeY* = 0), we choose the largest

[4] Note that ESS performs only one-dimensional tiling, and cannot incorporate two-dimensional tiling.

square tile for which the working set fits into the cache. $ws(x, y)$ gives the size of the working set for a tile with x rows and y columns. We maintain the number of columns to be a multiple of the cache line size to ensure that unnecessary data is not brought into the cache. If *ShapeX* and *ShapeY* have a non-zero value, the user wishes to guide the tile-selection process. *TypeX* and *TypeY* can take the value CONSTANT or VARIABLE. If *TypeX* (*TypeY*) has value CONSTANT and *TypeY* (*TypeX*) has value VARIABLE, as in cases 2 and 3, the user has fixed the number of rows (columns) in the tile as *ShapeX* (*ShapeY*). Procedure *SelectTile* only determines the number of columns (rows). If both *TypeX* and *TypeY* are CONSTANT (case 4), the user has already optimized the tile sizes, and the procedure performs no action. If both *TypeX* and *TypeY* are VARIABLE (case 5), the user intends the ratio of rows and columns of the tile to be *ShapeX* : *ShapeY*. The procedure selects the largest tile with the rows and columns in the given ratio for which the working set $\leq C$.

Set-associative caches can be effectively utilized in our tiling procedure to reduce cross interference in a tiled loop. We first determine the *largest allowed working set size* (M) for the tile by considering the possibility of cross-interference. If there is no cross-interference (e.g., there is only one array involved), we set $M = C$ (the cache size). However, there are circumstances where cross-interferences cannot be completely eliminated. For example, in the matrix multiplication benchmark implementing $P_1 \times P_2 = P_3$, where P_1, P_2, and P_3 are matrices, the tiling algorithm is designed to exploit reuse in matrix P_2 – the other matrices contribute unavoidable cross-interferences with the tile from P_2 in the cache. In this case, if the cache associativity $A > 1$, we can set $M = \frac{A-1}{A} \times S$. In other words, in an A-way cache, in any cache entry (with A lines), the *primary* tile (tile of the blocked array) occupies only $(A - 1)$ lines, leaving the remaining one line to be occupied by elements outside the reused tile. This ensures that elements of the primary tile are seldom evicted from the cache. After determining M, we use procedure *SelectTile* (Figure 4.22), replacing the condition $ws(x, y) \leq C$ by $ws(x, y) \leq M$.

4.3.3.2 Pad Size Computation. The computation of tile size above ignores the possibility of self-interferences within the tile. We now formulate the procedure for an appropriate padding of the rows of the array with dummy elements in order to avoid this self-interference. Consider the mapping of a 30 rows $\times$ 30 columns tile in a 256×256 array into a 1024-element cache. Figure 4.23(a) shows that rows 1 and 5 of the tile cause self-interference because they map into the same cache locations. To overcome this conflict, we can pad each row of the array with 8 dummy elements so that the 5th tile row now occupies the space adjacent to row 1 in the cache (Figure 4.23(b)). The regularity of the cache mapping ensures that no self-interference occurs among the elements of all the tile rows.

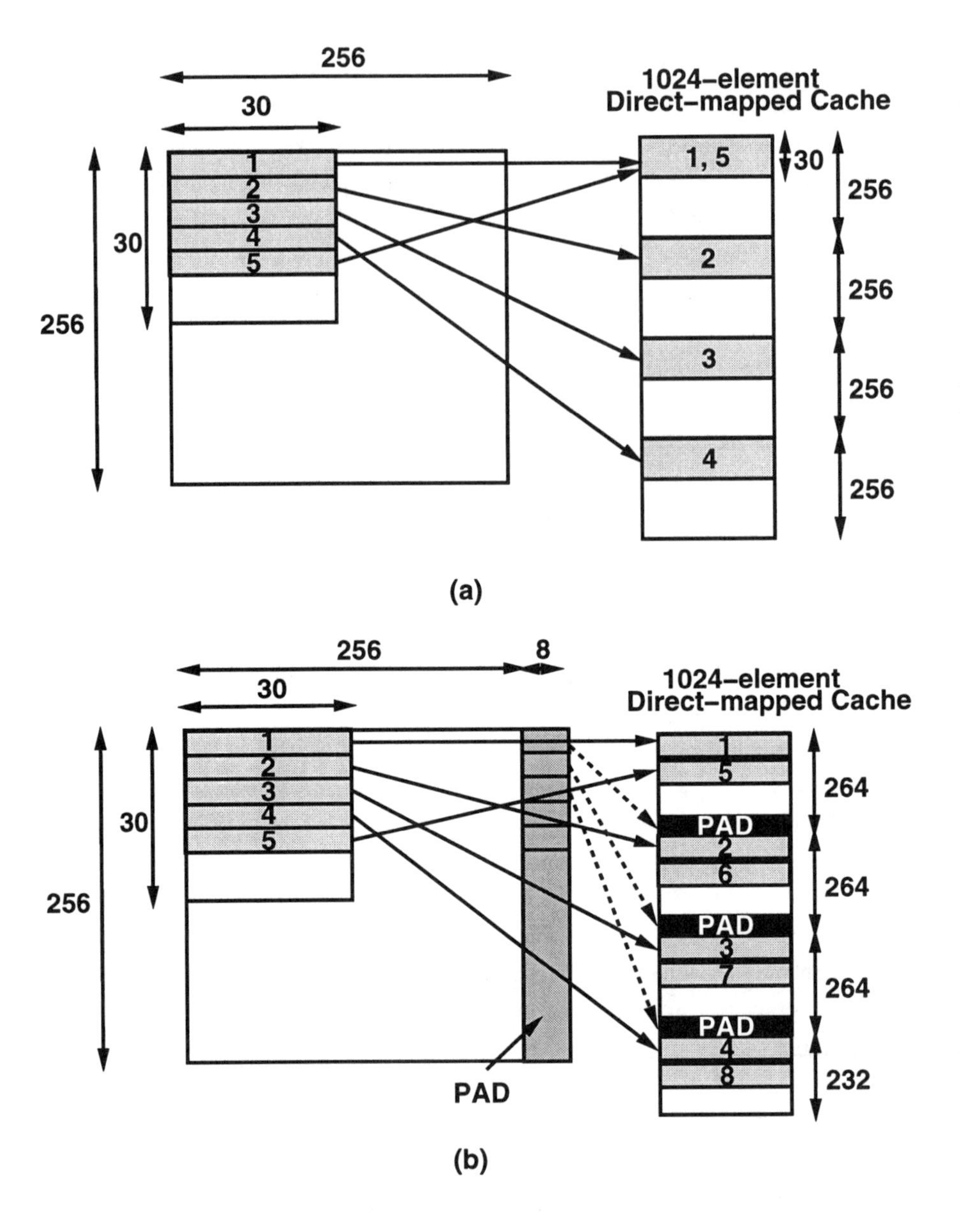

Figure 4.23. (a) Self-interference in a 30×30 tile: Rows 1 and 5 of tile interfere in cache (b) Avoiding self-interference by padding

Note that there could be more than one correct pad size that eliminates self-interference within a tile. Clearly, smaller pad sizes are better than larger ones because they reduce the wasted memory space. Algorithm *ComputePad* (Figure 4.24) outlines the pad size computation procedure.

```
Algorithm ComputePad
Input:
    Array Dimensions: N' × N; /* Row Size = N */
    Tile Dimensions: Rows × Cols
    Cache Size (in words): CacheSize
    Cache Line Size (in words): LineSize
Output:
    Pad Size for Array: PadSize

    if (N mod LineSize == 0) then InitPad = 0
    else InitPad = LineSize − N mod LineSize
    for PadSize = InitPad to CacheSize step LineSize
        Status = OK
        Initialize LinesArray[i] = 0 for all i
            (0 ≤ i ≤ ⌈CacheSize/LineSize⌉)
        for i = 0 to Rows
            for j = 0 to Cols step Linesize
                k = (i × (N + PadSize) + j) mod CacheSize
                if (LinesArray[k/LineSize] == 1)
                then Status = CONFLICT
                else LinesArray[k/LineSize] = 1
            end for /* loop j */
        end for /* loop i *
        if (Status == OK) then return PadSize
    end for /* loop PadSize */
end Algorithm
```

Figure 4.24. Algorithm for computing pad size for array

The inputs to algorithm *ComputePad* are: the cache line size, the cache size, and the dimensions of the array and the tile, and the output is *PadSize*, the smallest pad size that eliminates self-interference within the tile. The initial assignment to *InitPad* ensures that the rows of the padded array are aligned to the cache line size. For different multiples of *LineSize*, from a minimum of 0 to a maximum of *CacheSize*, we test if the resulting padded row of size $(N + PadSize)$ causes self-interference conflicts for the given tile of dimension $Rows \times Cols$. This is done by iterating through all the elements of the tile in steps of *LineSize*, and checking if any two elements map into the same cache line. In Figure 4.24, element (i, j) of the tile maps into cache location k, i.e., cache line number: $k / LineSize$. For every tile element (i, j), we update $LinesArray[k / LineSize]$ to 1 from 0. If $LinesArray[k / LineSize]$ is already 1, implying that at least one other tile element maps into this cache line, a conflict is flagged, and we repeat the procedure.

4.3.3.3 Multiple Tiled Arrays. The padding technique described above lends itself to an elegant generalization when computing the tile sizes of an algorithm involving more than one tiled arrays. We assume that the arrays have identical sizes ($S \times R$). Since both arrays are accessed in the same tiled loop, they have identical tile sizes. We choose the tile sizes such that, as before, the working set (which involves elements accessed from all the tiled arrays) fits into the cache. To determine the padding size such that both self-interferences within each tile, as well as cross-interferences across tiles are avoided, we first construct a tile consisting of the smallest rectangular shape (T) enclosing all the tiles. In Figure 4.25, the new contour formed from the tiles in arrays A and B is shown. We now determine the pad size such that rectangle T in an $S \times R'$ array does not incur self-interference, using the method described in Section 4.3.3.2. This becomes the pad size of all the tiled arrays. However, the tile size of each array remains $X \times Y$. Finally, we adjust the distance between the starting points of the arrays, so that the tiles are laid out the way they appear in rectangle T, in the first iteration. For example, in Figure 4.25, arrays A and B (both are now $S \times R'$) are laid out such that $A[0][0]$ and $B[0][0]$ are a distance $((X \cdot R') \bmod C)$ apart in the cache.

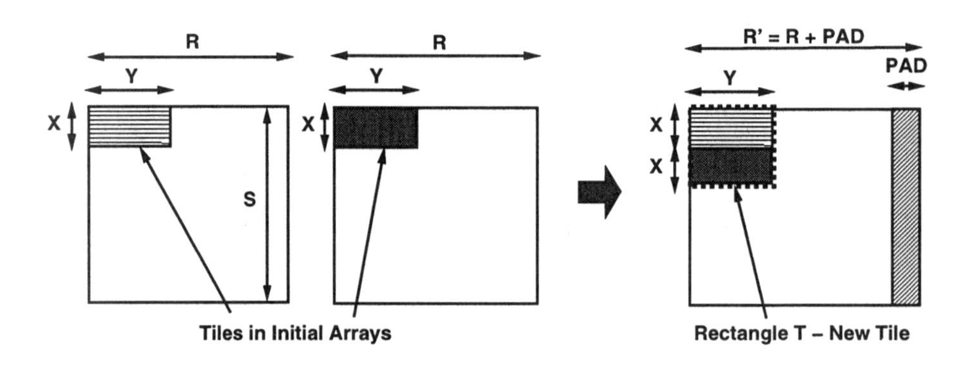

Figure 4.25. Multiple Tiled Arrays

4.3.4 Experiments

Experiments for the tiling strategy were performed on two common machines – SUN-Sparc5 (SS5) and SUNSparc10 (SS10). The SS5 has a direct-mapped 8 KB data cache (1024 double precision elements) with a 16-byte cache line (2 elements per line). The

SS10 has a 4-way associative 16 KB data cache (2048 elements) with a 32-byte cache line (4 double precision elements per line).

The example programs in the experiments are:

1. Matrix Multiplication (MM) (with the standard *ijk*-loop nest permuted to *ikj*-order, as in [LRW91]);

2. Successive Over Relaxation (SOR) [CM95];

3. L-U Decomposition without pivoting (LUD) [CM95]; and

4. Laplace [PTVF92].

We do not use the FFT algorithm in the comparisons, because it does not involve tiled loops, and consequently, LRW, TSS, and ESS cannot be applied to it.

The experimental results were obtained both through actual measurement (MFLOPs on SUN SPARC Stations) as well as simulations on the SHADE simulator from SUN Microsystems[Sun93]. Note that execution time is a function of the data cache miss ratio, the number of cache accesses, as well as the number of instructions executed. For tiled algorithms, the total number of instructions executed is smaller for larger tile sizes because the overheads introduced by tiling are smaller. The instruction cache misses are negligible for the above examples, as they all are small enough to fit into the instruction cache. We present performance comparisons of the DAT technique versus that of the LRW, ESS, and TSS algorithms in terms of both data cache miss ratio (misses per access) and MFLOPs.

4.3.4.1 Uniformity of Cache Performance. Our first experiment is to verify the claim that the padding technique results in a uniformly good performance for a wide variety of array sizes. Figure 4.26 shows the variation of data cache miss ratios (on the SS5 cache configuration) of four algorithms (LRW, ESS, TSS, and DAT) on the matrix multiplication (MM) example for all integral array sizes between 35 and 350.[5] We observe that the miss ratio of DAT is consistently low, independent of problem size, whereas all other algorithms show some sensitivity to the size. This is attributed to the fact that DAT uses fixed tile dimensions (30×30) for the given cache parameters, independent of array size, whereas in the other algorithms, tile dimensions vary widely with array size. Note that although the *number of misses*, and *number of instruction executed*, jointly determine the actual performance, we have plotted the miss ratio here just to show that the miss ratios are consistently low for DAT (low

[5]For small array sizes, all the data fits into the cache, obviating the necessity for tiling; for larger sizes, the simulation times on the commercial simulator, SHADE, were too long to examine every integral data size. Hence the range 35-350.

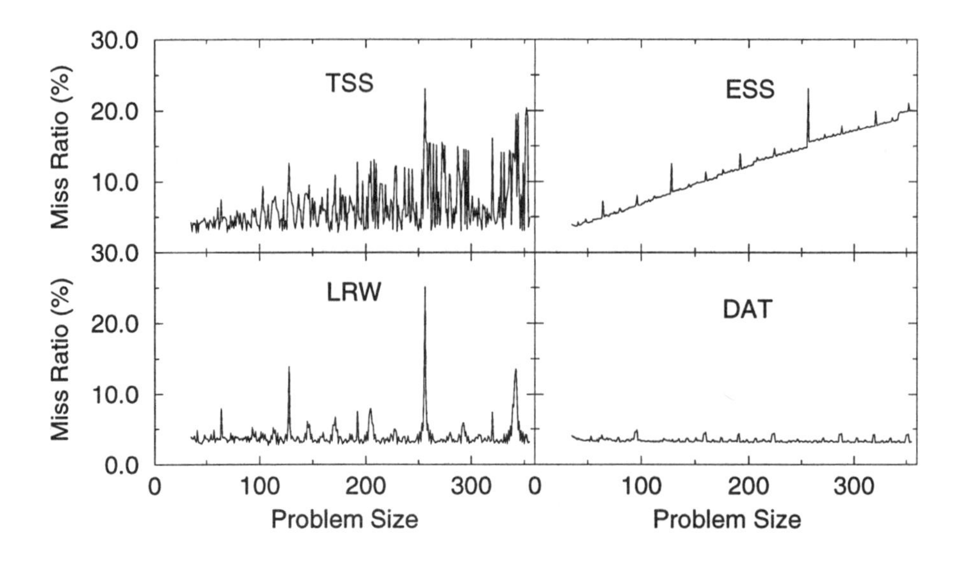

Figure 4.26. Matrix multiplication on SS-5: Variation of data cache miss ratio with array dimension

miss ratio is necessary, but not sufficient). Although LRW also has a low miss ratio on an average, it often selects small tile sizes leading to an increase in the number of instructions, as described later. Figure 4.26 demonstrates that DAT exhibits stability of cache performance across the range $35 - 350$ of problem sizes. We expect similar performance for larger array sizes.

4.3.4.2 Variation of Cache Performance with Problem Size. We present experiments below on the performance of each technique on the various examples, for several array sizes. We include array sizes of 256, 300, 301, and 550 to enable comparison with TSS [CM95] (which also presents data on these sizes). Array sizes 256 and 512 are chosen to illustrate the case with pathological cache interference, while 300 and 301 are chosen to demonstrate the widely different tile sizes for small changes in array size. Sizes 384, 640, and 768 illustrate the cases where the array size is a small multiple of a power of 2 ($384 = 128 \times 3; 640 = 128 \times 5; 768 = 256 \times 3$). We also include data for array sizes 700, 800, 900 and 1000.

For each example, we report results of execution on the SS5 and SS10 workstations. The SS5 has an 8KB direct-mapped cache, whereas the SS10 has a 16KB 4-way associative cache, with no second level cache. The results of our experiments on these two platform illustrate the soundness of our technique on direct-mapped as well as associative caches. We present the variation of the following parameters with respect to array size.

Miss Ratio – The data cache miss ratio (misses per data cache access).

Instructions – The total number of instructions executed, normalized to the instruction count of our technique DAT (i.e., value for DAT is 1.0, and that for other algorithms is divided by the instruction count of DAT).

Cycles – The approximate number of processor cycles, assuming a memory latency of 12 cycles, normalized to the value for DAT. We use an approximate formula:

$$\#\text{Cycles} = \text{Instruction Count} + \text{Latency} \times \text{Data Cache Misses}$$

This formula is a good approximation for cycle count in a single-instruction stream processor, and takes into account both data cache misses as well as the number of instructions. As explained earlier, the instruction cache misses are negligible.

MFLOPs – The measured MFLOPs rate.

In the graphs shown in Figure 4.27, the above parameters are along the y-axes of the graphs, whereas the x-axis of all the graphs is the array size (i.e., the problem size).

4.3.4.3 L-U Decomposition. Figure 4.27 shows the cache performance of the blocked L-U Decomposition program described in [CM95]. An analysis of the array access patterns in LUD reveals the following relationship for the optimal tile shape:

$$\text{No. of Columns} = \text{No. of Rows} \times \text{Cache Line Size} \tag{4.2}$$

The analysis is described in Appendix 4.A.

Figure 4.27 shows that DAT and LRW have the lowest cache **miss ratios** (except at array sizes such as 256, 384, 512, etc., where LRW has considerably higher miss ratios). The lower miss ratio of LRW is, however, often at the expense of instruction count, since the relatively smaller tiles in LRW incur overheads introduced by loop tiling. TSS tends to incur higher miss ratios because it sometimes chooses a comparatively larger tile sizes, leading to the working set size being too close to the cache size, which results in cross-interferences. ESS incurs cache conflicts because it does not account for cross-interferences.

A comparison of the **instruction count** shows that, in general, ESS has a smaller number of executed instructions than DAT and LRW. This is a consequence of the larger tile sizes selected by ESS. The instruction count of TSS fluctuates because it selects widely varying tile sizes.

The processor **cycle count** (with memory latency = 12 cycles) shows that DAT has the best overall performance. It is interesting to note that the curves follow roughly

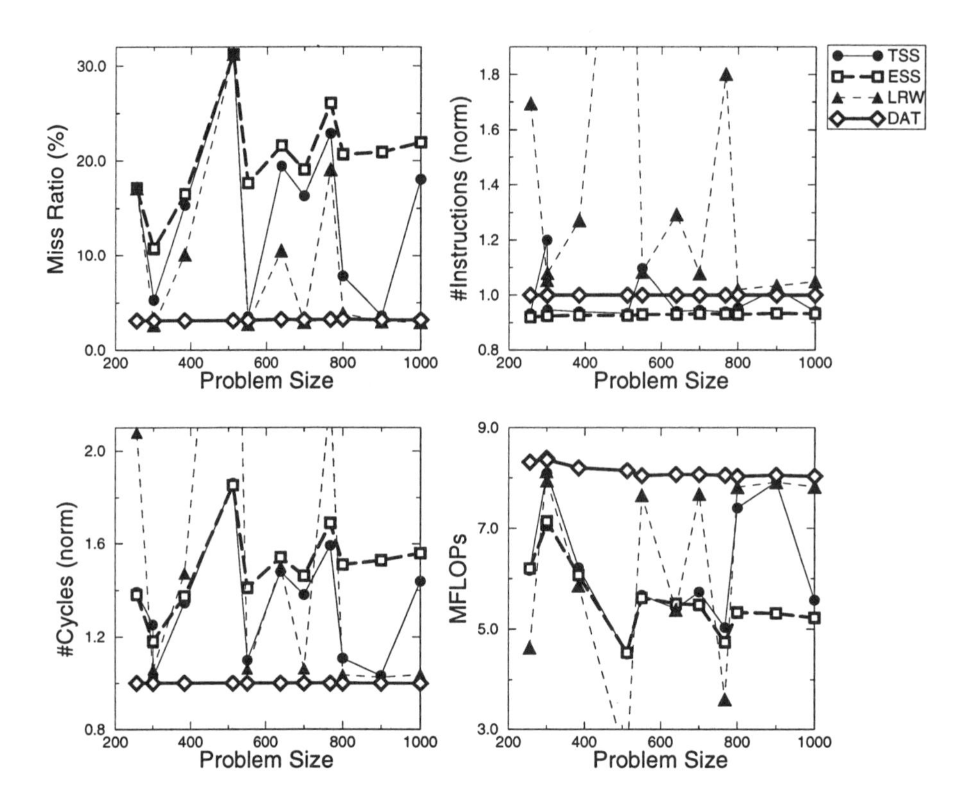

Figure 4.27. Performance of L-U Decomposition on SS-5: Variation of (1) Data Cache Miss Ratio, (2) Normalized Instruction Count, (3) Normalized Processor Cycles, and (4) MFLOPs with array size

the same shape as the cache miss ratio curves, indicating that cache miss ratio is a very important factor in determining performance.

The **MFLOPs** measurements show DAT to perform better than other algorithms. The MFLOPs graph is, actually, the inverse of the cycle count graph, with the difference that the former is a measurement of execution time on the machine, whereas the latter is a computation from the simulation results. The close correlation of the two graphs indicates that the cycle count computations are reliable.

In summary, Figure 4.27 shows that our technique DAT consistently incurs lower miss ratios and exhibits better performance than previous techniques for the L-U decomposition example. We observe similar performance results in our experiments using the LUD benchmark on the SS10 machine.

Matrix Multiplication

The results for *matrix multiplication* is summarized in Table 4.3. A comparison of the total number of instructions executed showed that, in general, a larger tile size implies lesser number of instructions due to reduced overheads. We assumed the tile size 30×30 for the SS-5 cache (direct mapped, 1024-element), and 36×36 for SS-10 cache (4-way set-associative, 2048-element – we used the formulation for A-way associative cache discussed in Section 4.3.3.1). Note that we only consider multiples of cache line size of candidates for tile size. Here we assumed that no information was available about the characteristic of the application and hence used a square tile.

Successive Over Relaxation (SOR)

The results for *SOR* is summarized in Table 4.3. In this case, an analysis reveals that maximal reuse is obtained by choosing the longest (i.e., maximum number of columns) tile that avoids self-interference. We observed that ESS (which selects as many rows as will fit into the cache) has the same performance as DAT for the smaller array sizes, because both techniques generate the same tile sizes. However, for sizes 550 and greater, DAT achieves better performance because of better reuse.

Laplace

The results for *Laplace* is summarized in Table 4.3. The performance of DAT is better than the other algorithms because the Laplace kernel involves the blocking of multiple arrays – a feature handled by DAT, but not the other techniques. DAT is able to avoid all cross-interference between the arrays, whereas TSS, ESS, and LRW all incur cross-interference.

4.3.4.4 Summary of Results. Table 4.3 summarizes the comparisons of experimental results for all the four examples (MM, LUD, SOR, and Laplace) on the SS5 and SS10 platforms, on the basis of the four metrics identified above. In order to do a fair comparison, the table does not include array sizes such as 512, 640, etc., which would penalize all the other techniques. Thus, the improvements shown in the table are obtained without considering these cases. If these were taken into account, the improvement would be higher. Column 1 gives the example names; Columns 2, 3, and 4 give the improvement in cache miss ratios (MR) of our technique DAT (D) over TSS (T), ESS (E), and LRW (L) respectively. Similarly, Columns 5, 6, and 7 compare the normalized instruction counts; Columns 8, 9, and 10 compare the normalized processor cycle counts; and Columns 11, 12, and 13 compare the MFLOPs with respect to DAT, i.e., the speedup of DAT over other techniques. On the SS5, we notice an average speedup of 24.2% over TSS, 38.0% over ESS, and 12.0% over LRW. On the SS10, we observe an average speedup of 7.2% over TSS, 13.8% over ESS, and 4.8% over LRW. The speedup on SS10 is smaller because the SS10 has a larger cache, which reduces the number of cache conflicts.

Table 4.3. Summary of performance results on SS5 and SS10. Data for problem sizes 384, 512, 640, and 768 are not included because that would penalize LRW (L), ESS (E), and TSS (T); the cache performance behavior for all examples is similar to that observed in Figure 4.26

SUN SPARC 5												
Ex.	MR(x)(%) – MR(D)(%)			Inst(x)/ Inst(D)			Cycle(x)/ Cycle(D)			Speedup (D/x)		
	T	E	L	T	E	L	T	E	L	T	E	L
MM	10.21	22.03	3.76	0.99	0.95	1.07	1.29	1.61	1.19	1.30	1.64	1.14
LUD	6.45	14.18	1.59	1.00	0.93	1.14	1.22	1.40	1.17	1.22	1.40	1.13
SOR	8.81	10.28	0.79	1.03	1.00	1.10	1.31	1.34	1.10	1.23	1.26	1.07
Lap	6.54	6.22	0.35	1.00	1.00	1.07	1.17	1.16	1.06	1.22	1.22	1.14
SUN SPARC 10												
Ex.	MR(x)(%) – MR(D)(%)			Inst(x)/ Inst(D)			Cycle(x)/ Cycle(D)			Speedup (D/x)		
	T	E	L	T	E	L	T	E	L	T	E	L
MM	4.01	9.35	1.39	1.00	0.94	1.03	1.15	1.29	1.08	1.15	1.30	1.06
LUD	2.47	4.54	1.06	0.99	0.96	1.08	1.08	1.13	1.12	1.08	1.13	1.09
SOR	0.72	2.09	1.20	1.05	1.00	1.05	1.07	1.09	1.09	1.02	1.05	1.00
Lap	1.27	1.88	0.03	1.02	1.00	1.04	1.05	1.05	1.03	1.04	1.07	1.04

4.3.5 Discussion

The main advantage of DAT, the data alignment technique we presented, is its flexibility – decoupling tile selection from the padding phase allows the tile size to be independently optimized without regard to self-interference conflicts. This allows us to select larger tile sizes to maximize the cache utilization. This also allows us to incorporate a user-supplied tile-shape which might be optimized for a specific application. For instance, we can easily handle one-dimensional tiling, while it cannot be easily incorporated into TSS and LRW.

For a given application and cache size, we choose a fixed tile size, independent of the array size (only the pad size varies with the array size). An important consequence of this choice is the stability of the performance of the resulting tiled loop, in comparison to TSS, ESS, and LRW (all of which select tile sizes that are very sensitive to the array size). For instance, in the matrix multiplication example shown in Figure 4.27, the miss ratio of DAT was in the range [3.11% – 4.87%], in comparison to [2.77% – 25.18%] for LRW, [2.93% – 23.18%] for TSS, and [3.70% – 23.15%] for ESS. Further, Figure 4.27 shows that all the other techniques suffer from pathological interference, not only for array sizes that are powers of two (256 and 512), but also for small multiples of powers of two (384, 640, and 768). It should be noted, however, that these array sizes are not uncommon in applications. For cache line sizes greater than 1, we observe the same behavior when the sizes are close to one of these numbers, e.g., 254, 255, 385, etc.

Another advantage of DAT is that it is possible to avoid cross-interferences among several tiled arrays, while all the other techniques – LRW, TSS, and ESS are targetted at algorithms in which the reuse of only a single array dominates. While accesses to the several arrays would cause cross-interference in the other approaches, DAT is able to completely eliminate cache conflicts arising out of this cross-interference using our technique.

Finally, DAT can be easily extended to arrays with greater than two dimensions, by serializing the padding procedure over the different dimensions.

One consequence of the padding strategy is that the structure of the data arrays is transformed, and references to the array in the rest of the code has to reflect the new structure. This, however, does not lead to any performance penalties. For example, when the row size is updated from R to R', the expression to compute the location for $a[i][j]$ changes from $(A + i \times R + j)$ to $(A + i \times R' + j)$, assuming the array begins at A.

There could be a conflict in padding sizes if the same array is accessed in different tiled loops. In this case, we invoke the padding strategy on the more critical loop, and in the others, use one of the existing methods (LRW or TSS) in the remaining loops, with updated row size for the array. This strategy works because LRW and TSS, which do not modify the data structure, can now use the size of the modified array to generate the tile sizes. However, an examination of existing benchmark programs reveals that such a situation (same array being accessed in two loop nests with different tiling patterns) is rare. It is well known that in most memory-intensive programs, most of the computation tends to be concentrated in a few important kernels.

4.4 SUMMARY

Code generation for embedded processors reveals the scope for many optimizations that have been hitherto unaddressed in traditional compilers. An important set of features that can be exploited while generating code for embedded processors is the parameters of the data cache. In this chapter, we demonstrated how a careful data layout strategy for scalar and array variables that takes into account the parameters of the data cache, such as cache line size and cache size, could effect significant performance improvements in the execution of embedded code. We then presented an extension of the data alignment strategy to solve the tile size selection problem in the tiling/blocking compiler optimization. Data alignment through an appropriate padding of two-dimensional arrays leads to consistently good cache performance, independent of the array size. The blocking optimization is useful not only for scientific code, but also in the context of embedded applications. For example, image processing algorithms that compute the value of a pixel using its value and that of its neighbours, have identical memory access patterns to the SOR algorithm discussed in Section 4.3.4.

Appendix 4.A: Tile Shape Computation in L-U Decomposition Example

We present here the procedure for computation of the tile-shape for the L-U Decomposition problem.

Let the matrix size be $M \times M$.

Let L = Cache Line Size.

Let k be the outer loop index and K be the range over which k iterates.

For a tile with a given number of elements (N), we wish to find the number of rows (r) and columns (c) for the tile ($N = r \times c$), which will minimise the number of cache misses.

Total number of cache misses for a tile $= K \times (r + \frac{c}{L})$, since in each iteration, one row of the tile causes $\frac{c}{L}$ misses and one column causes r misses. Note that K is a constant independent of the tile size.

Total number of tiles $\approx (\frac{M}{r} \times \frac{M}{c})$ (approximation for $\lceil \frac{M}{r} \rceil \times \lceil \frac{M}{c} \rceil$).

Hence, the total number of misses $\approx K \times (\frac{M}{r} \times \frac{M}{c}) \times (r + \frac{c}{L}) \approx K' \times (\frac{1}{c} + \frac{1}{rL})$

Now, since $r \times c = N$, and $c \times rL = NL =$ *constant*, we note that the number of misses is minimised for $c = rL$ (using the fact that for two numbers a and b with constant product ab, the sum $\frac{1}{a} + \frac{1}{b}$ is minimised for $a = b$).

In other words, for minimising the number of cache misses, we have:

$$\text{No. of Columns} = \text{No. of Rows} \times \text{Cache Line Size} \tag{4.A.1}$$

5 ON-CHIP VS. OFF-CHIP MEMORY: UTILIZING SCRATCH-PAD MEMORY

5.1 SCRATCH-PAD MEMORY

In addition to a data cache that interfaces with slower off-chip memory, a fast on-chip SRAM is often used in several applications, so that critical data can be stored there with a guaranteed fast access time. The on-chip SRAM, termed **Scratch-Pad memory**, refers to data memory residing on-chip, that is mapped into an address space disjoint from the off-chip memory, but connected to the same address and data buses. Both the cache and Scratch-Pad SRAM allow fast access to their residing data, whereas an access to the off-chip memory (usually DRAM) requires relatively longer access times. The main difference between the Scratch-Pad SRAM and data cache is that, the SRAM guarantees a single-cycle access time, whereas an access to the cache is subject to compulsory, capacity, and conflict misses.

The concept of Scratch-Pad memory is an important architectural consideration in modern embedded systems-on-chip, where advances in fabrication technology have made it possible to combine DRAM and ordinary logic on the same chip. As a result, it is possible to incorporate embedded DRAMs along with a processor core in the same chip [Mar97, Wil97]. Since data stored in embedded DRAM can be accessed much faster than that in off-chip DRAM, a related optimization problem that arises in this context is how to identify critical data in an application for storage in on-chip memory. We use the terminology *Scratch-Pad SRAM* to include the embedded DRAM configuration.

In this chapter, we address the issue of minimization of the total execution time of an embedded application by a careful partitioning of scalar and array variables used in the application into off-chip DRAM (accessed through data cache) and Scratch-Pad SRAM. Figure 5.1 highlights the components of the embedded system architecture addressed by the techniques presented in this chapter.

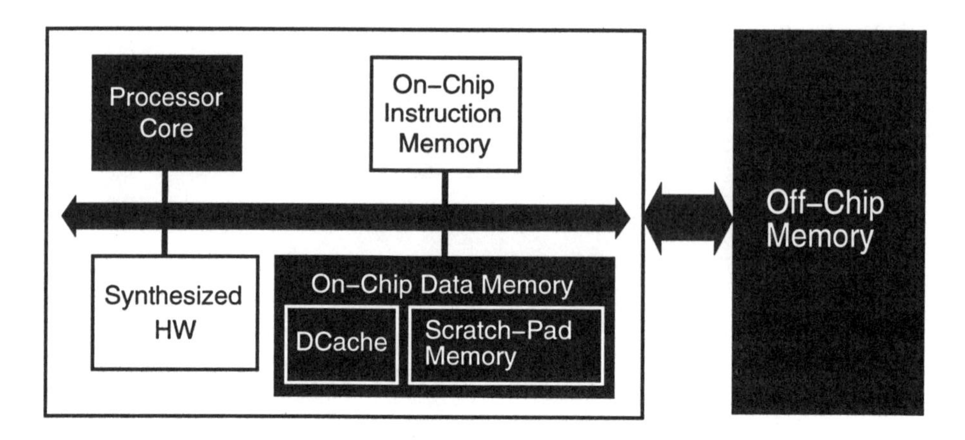

Figure 5.1. Shaded blocks are targetted by the optimization techniques presented in this chapter

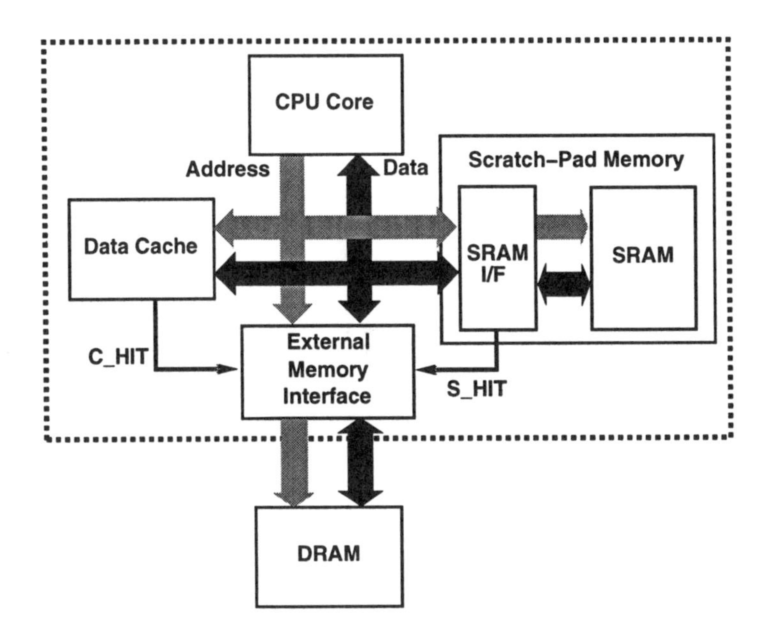

Figure 5.2. Block Diagram of Typical Embedded Processor Configuration

5.2 PROBLEM DESCRIPTION

Figure 5.2 shows the architectural block diagram of an application employing an embedded core processor with Scratch-Pad memory. The address and data buses from the CPU core connect to the Data Cache, Scratch-Pad memory, and the *External Memory Interface* (EMI) blocks. On a memory access request from the CPU, the data cache indicates a cache hit to the EMI block through the **C_HIT** signal. Similarly, if the SRAM interface circuitry in the Scratch-Pad memory determines that the referenced memory address maps into the on-chip SRAM, it assumes control of the data bus and indicates this status to the EMI through signal **S_HIT**. If both the cache and SRAM report misses, the EMI transfers a block of data of the appropriate size (equal to the cache line size) between the cache and the DRAM.

The data address space mapping is shown in Figure 5.3, for a sample addressable memory of size N data words. Memory addresses $0 \ldots P-1$ map into the on-chip Scratch-Pad memory, and have a single processor cycle access time. Thus, in Figure 5.2, S_HIT would be asserted whenever the processor attempts to access any address in the range $0 \ldots P-1$. Memory addresses $P \ldots N-1$ map into the off-chip DRAM, and are accessed by the CPU through the data cache. A cache hit for an address in the range $P \ldots N-1$ results in a single-cycle delay, whereas a cache miss, which leads to a block transfer between off-chip and cache memory, results in a delay of 10-20 processor cycles.

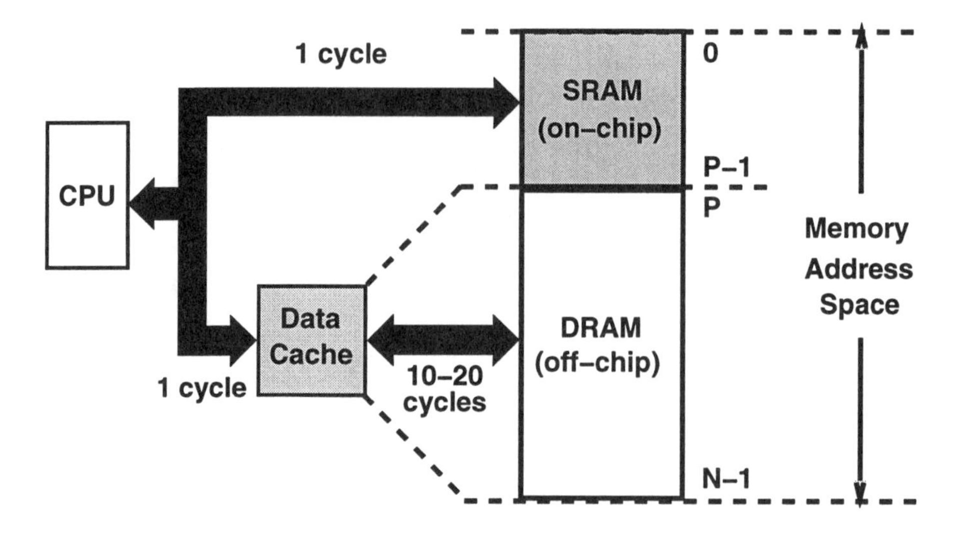

Figure 5.3. Division of Data Address Space between SRAM and DRAM

We assume that register allocation has already been performed. Given the embedded application code, our goal is to determine the mapping of each scalar and arrayed program variable into local Scratch-Pad SRAM or off-chip DRAM, while maximizing the application's overall memory access performance.

The sizes of the data cache and the Scratch-Pad SRAM are limited by the total area available on-chip, as well as by the single cycle access time constraint. Hence, we must first justify the need for both the data cache and SRAM. We motivate the need for both types of on-chip memory using the following example. Suppose the embedded core processor in Figure 5.2 can support a total of 2 KBytes for the data cache and the SRAM. We can analyze the pros and cons of four extreme configurations of the cache and SRAM:

(1) No local memory: In this case, we have the CPU accessing off-chip memory directly, and spending 10–20 processor cycles on every access. Data locality is not exploited and the performance is clearly inferior in most cases.

(2) Scratch-Pad memory of size 2K: In this case, we have an on-chip SRAM of larger size, but no cache. The CPU has an interface both to the SRAM and the off-chip memory. When large arrays that do not fit into the SRAM are present, the direct interface to external memory has to be used, thereby degrading performance.

(3) Data cache of size 2K: Here, we have a larger data cache, but no separate local SRAM. In many cases, having only a cache results in certain unavoidable cache misses that degrade performance due to stalled CPU cycles. We illustrate this with the example of a *Histogram Evaluation* code shown in Figure 5.4 from a typical

Image Processing application, which builds a histogram of 256 brightness levels for the pixels of an 512×512 image.

```
char BrightnessLevel [512][512];
int Hist [256];        /* Elements initialized to 0 */
...
for (i = 0; i < 512; i++)
    for (j = 0; j < 512; j++)
        /* For each pixel (i, j) in image */
        level = BrightnessLevel [i][j];
        Hist [level] = Hist [level] + 1;
    end for
end for
```

Figure 5.4. Histogram Evaluation Example

The performance is degraded by the conflict misses in the cache between elements of the two arrays *Hist* and *BrightnessLevel*. Data layout techniques (Chapter 4) are not effective in eliminating the above type of conflicts, because the accesses to *Hist* are data-dependent.

(4) 1K Data cache + 1K Scratch-Pad SRAM: The problem incurred in (3) above could be solved elegantly using the architecture of Figure 5.2, with a 1K data cache and 1K SRAM. Since the *Hist* array is relatively small, we can store it in the SRAM, so that it does not conflict with *BrightnessLevel* in the data cache. This storage assignment improves the performance of the *Histogram Evaluation* code significantly. The single-cycle access time guarantee for data stored in SRAM and the possibility of avoiding conflicts makes the architecture with a combination of cache and Scratch-Pad memory superior to cache alone.

From the above, it is clear that both the SRAM and data cache are desirable. Note that there could be applications where the Scratch-Pad SRAM offers no particular advantage over a single cache – for example, when all arrays are too big to fit into SRAM; or there is little temporal reuse among the arrays so that the cache conflicts do not cause any performance penalty. However, experiments show that the SRAM improves performance in most typical applications.

Next, we address the problem of partitioning scalar and array variables in an application code into Scratch-Pad memory and off-chip DRAM accessed through data cache, to maximize the performance by selectively mapping to the SRAM those variables that are estimated to cause the maximum number of conflicts in the data cache. We assume that the array sizes and loop bounds are known, either statically, or through profiling. This is a reasonable assumption for many embedded applications.

5.3 THE PARTITIONING STRATEGY

The overall approach in partitioning program variables into Scratch-Pad memory and DRAM is to minimize the interference between different variables in the data cache. Let us first outline the different features of the code affecting the partitioning, and then present a partitioning strategy based on these features.

5.3.1 Features Affecting Partitioning

The partitioning of variables is governed by the following code characteristics:

1. scalar variables and constants;
2. size of arrays;
3. life-times of variables;
4. access frequency of variables; and
5. conflicts in loops.

5.3.1.1 Scalar Variables and Constants. In order to prevent interference with arrays in the data cache, we map all scalar variables and constants to the Scratch-Pad memory. This assignment helps avoid the kind of conflicts mentioned in Section 5.2. If scalars are mapped to the DRAM, (and, consequently, accessed through the cache), it may be impossible to avoid cache conflicts with arrays, because arrays are assigned to contiguous blocks of memory, parts of which will map into the same cache line as the scalars, causing conflict misses.

It is possible to do a more sophisticated analysis of the most frequently accessed scalars to the SRAM, but our decision to map all scalars to the SRAM is based on our observation that for most applications, the memory space attributable to scalars is negligible compared to that occupied by arrays.

5.3.1.2 Size of Arrays. We map arrays that are larger than the SRAM into off-chip memory, so that these arrays are accessed through the data cache. Mapping large arrays to the cache is the natural choice, as it simplifies the array addressing. If a part of the array were to map into the SRAM, the compiler would have to generate book-keeping code that keeps track of which region of the array is addressed, thereby making the code inefficient. Further, since loops typically access array elements more or less uniformly, there is little or no motivation to map different parts of the same array to memories with different characteristics.

5.3.1.3 Life-times of Variables. The *life-time* of a variable, defined as the period between its *definition* and *last use* [ASU93], is an important metric affecting register allocation. Variables with disjoint lifetimes can be stored in the same processor register. The same analysis, when applied to arrays, allows different arrays to share the same memory space.

The life-time information can also be used to avoid possible conflicts among arrays. To simplify the conflict analysis, we assume that an array variable accessed inside a loop is *alive* throughout the loop. Figure 5.5 shows an example lifetime distribution of four array variables, a, b, c, and d. Since b and c have disjoint life-times, the memory space allocated to array b can be reused for array c. Since arrays a and d have intersecting life-times, cache conflicts between them can be avoided by mapping one of them to the SRAM. However, since on-chip resources are scarce, we need to determine which array is more critical. The quantification of the criticality depends on the access frequency of the variables.

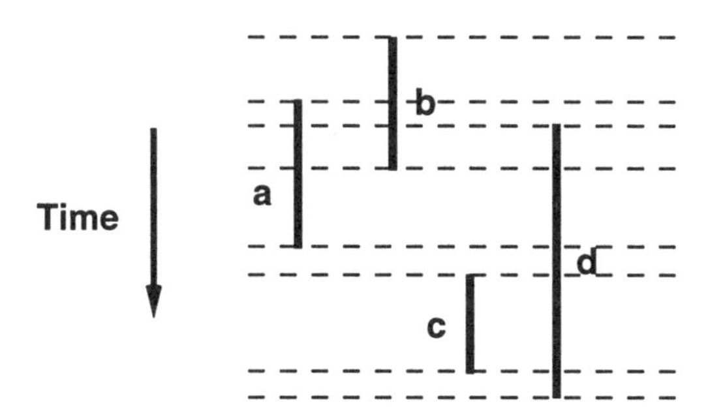

Figure 5.5. Example Life-Time Distribution

5.3.1.4 Access Frequency of Variables. To obtain an estimate of the extent of conflicts, we have to consider the frequency of accesses. For example, in Figure 5.5, note that array d could potentially conflict with all the other three arrays, so we could consider storing d in the SRAM. However, if the number of accesses to d is relatively small, it is worth considering some other array (e.g., a) first for inclusion in SRAM, because d does not play a significant role in cache conflicts. For each variable u, we define the *Variable Access Count, VAC(u)*, to be the number of accesses to elements of u during its lifetime.

Similarly, the number of accesses to *other* variables during the lifetime of a variable, is an equally important determinant of cache conflicts. For each variable u, we define the *Interference Access Count, IAC*(u), to be the number of accesses to other variables during the lifetime of u.

We note that each of the factors discussed above, *VAC*(u) and *IAC*(u), taken individually, could give a misleading idea about the possible conflicts involving variable

u. Clearly, the conflicts are determined jointly by the two factors considered together. A good indicator of the conflicts involving array u is given by the sum of the two metrics. We define the *Interference Factor, IF,* of a variable u as:

$$IF(u) = VAC(u) + IAC(u) \tag{5.1}$$

A high *IF* value for u indicates that u is likely to be involved in a large number of cache conflicts if mapped to DRAM. Hence, we choose to map variables with high *IF* values into the SRAM.

5.3.1.5 Conflicts in Loops. The *IF* factor defined in the previous section can be used to estimate conflicts in straight-line code and conditionals (i.e., non-loop code). In the case of arrays accessed in loops, it is possible to make a finer distinction based on the array access patterns. Consider a section of a code in which three arrays a, b, and c are accessed, as shown in Figure 5.6(a).

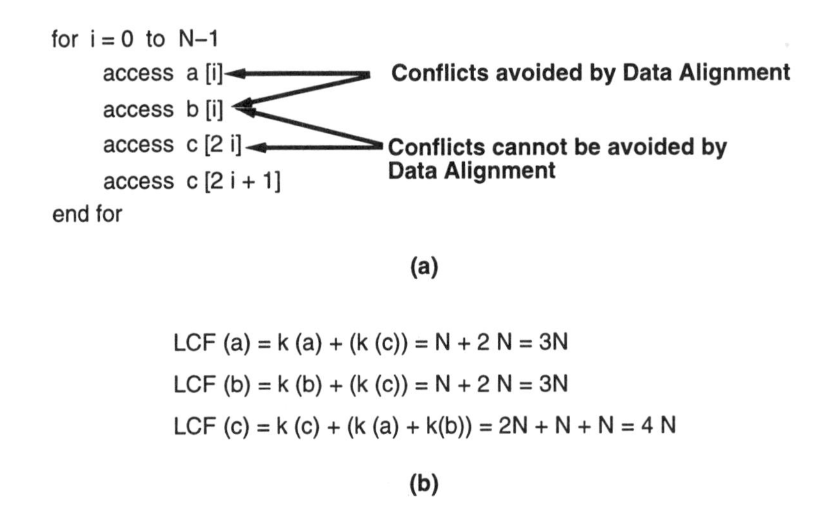

Figure 5.6. (a) Example Loop (b) Computation of LCF values

We notice that arrays a and b have an identical access pattern, which is different from that of c. Data alignment techniques (Chapter 4) can be used to avoid data cache conflicts between a and b – the arrays can be appropriately displaced in memory so that they never conflict in the cache. However, when the access patterns are different, cache conflicts are unavoidable (e.g., between b and c – the access patterns are different due to differing co-efficients of the loop variables in their index expressions). In such circumstances, conflicts can be minimized by mapping one of the conflicting arrays to the SRAM. For instance, conflicts can be eliminated in the example above, by mapping a and b to the DRAM/cache, and c to the Scratch-Pad memory.

To accomplish this, we define the *Loop Conflict Factor, LCF* for a variable u as:

$$LCF(u) = \sum_{i=1}^{p}(k(u) + \sum_{v} k(v)) \tag{5.2}$$

where the summation $\sum_{i=1}^{p}$ is over all loops $(1 \ldots p)$ in which u is accessed, and $\sum_{v}$ is over all arrays (other than u) that are also accessed in loop i, and for which it is not possible to use data placement techniques to completely eliminate cache conflicts with u. In the example above, where we have only one loop ($p = 1$), the LCF values shown in Figure 5.6(b) are generated. We have one access to a and two to c in one iteration of the loop. Total number of accesses to a and c combined is: $N + 2N = 3N$. Thus, we have $LCF(a) = 3N$, since cache conflicts between a and b can be completely eliminated by data placement techniques. Similarly, $LCF(b) = 3N$, and $LCF(c) = 4N$. The *LCF* value gives us a metric to compare the criticality of loop conflicts for all the arrays. In general, the higher the *LCF* number, the more conflicts are likely for an array, and hence, the more desirable it is to map the array to the Scratch-Pad memory.

5.3.2 Formulation of the Partitioning Problem

In Sections 5.3.1.4 and 5.3.1.5, we defined two factors, Interference Factor (*IF*) and Loop Conflict Factor (*LCF*). We integrate these two factors and arrive at an estimate of the total number of accesses that could theoretically lead to conflicts due to an array. We define the *total conflict factor (TCF)* for an array u as:

$$TCF(u) = IF(u) + LCF(u) \tag{5.3}$$

where the $IF(u)$ value is computed over a lifetime that excludes loops. *TCF(u)* gives an indication of the total number of accesses that are exposed to cache conflicts involving array u, and hence, denotes the importance of mapping u to the SRAM.

We formulate the data partitioning problem as follows:

Given a set of n arrays $A_1 \ldots A_n$, with TCF values $TCF_1 \ldots TCF_n$; sizes $S_1 \ldots S_n$; and an SRAM of size S, find an optimal subset $Q \subseteq \{1, 2, \ldots, n\}$ such that $\sum_{i \in Q} S_i \leq S$, and $\sum_{i \in Q} TCF_i$ is maximized.

We note that the problem, as formulated above, is similar to the *Knapsack Problem* [GJ79]. However, there is an additional factor to consider in this case – *several arrays with non-intersecting lifetimes can share the same SRAM space.*

5.3.3 Solution of the Partitioning Problem

An exhaustive-search algorithm to solve the memory assignment problem would have to first generate clusters of all combinations of *compatible arrays* (arrays that can share the same SRAM space) and then generate all possible combinations of these clusters and pick the combination with total size fitting into the SRAM that maximizes the cumulative TCF value. This procedure requires $O(2^{2^n})$ time, which is unacceptably expensive, since the function $y = 2^{2^n}$ grows very rapidly, even for small values of n.

In our solution to the memory data partitioning problem, we first group arrays that could share SRAM space into clusters, and then use a variation of the *value-density* approximation algorithm [GJ79] for the Knapsack Problem to assign clusters to the SRAM. The approximation algorithm first sorts all the items in terms of *value per weight* (i.e., the *value density*), and selects items in decreasing order of value-density until no more items can be accommodated. We define *Access Density (AD)* of a cluster c as:

$$AD(c) = \frac{\sum_{v \in c} TCF(v)}{\max\{\text{size}(v) | v \in c\}} \tag{5.4}$$

and use this factor to assign clusters of arrays into the SRAM. Note that the denominator is the size of the largest array in the cluster. This is because, at any time, only one of the arrays is *live*, and consequently, the arrays, which share the same memory space, need to be assigned only as much memory as the largest in the cluster.

The memory data partitioning algorithm *MemoryAssign* is shown in Figure 5.7. The input to algorithm is the SRAM size and the application program P, with the register-allocated variables marked. The output is the assignment of each variable to Scratch-Pad memory or DRAM.

The algorithm first assigns the scalar constants and variables to the SRAM, and the arrays that are larger than the Scratch-Pad memory, to the DRAM. For the remaining arrays, we first generate the *CompatibilityGraph G*, in which the nodes represent arrays in program P, and an edge exists between two nodes if the corresponding arrays have disjoint life-times. The analogous problem of mapping scalar variables into a register file is solved by *clique partitioning* [GDLW92] of the compatibility graph. However, we cannot apply clique partitioning algorithm in a straightforward manner when we cluster arrays, because *the cliques we are interested in, are not necessarily disjoint.* For example, consider a clique consisting of three arrays, A, B, and C, where size(A) is larger than size(B) and size(C). Assuming that the currently available SRAM space during one iteration of the assignment is *AvSpace*, the clique $\{A, B, C\}$ will not fit into the SRAM if size(A) > *AvSpace*. However, the subset $\{B, C\}$ will fit into the SRAM if size(B) and size(C) are smaller than *AvSpace*. Thus, we need to consider both cliques $\{A, B, C\}$ and $\{B, C\}$ during the SRAM assignment phase.

To handle the possibility of overlapping clusters, we generate one cluster (clique in graph G) $c(u)$ for every array u, which consists of u and all nodes v in graph G, such that size(v) $\leq$ size(u), and the subgraph consisting of the nodes of $c(u)$ is fully connected. Thus, for each array u, we attempt to select the clique with maximum

Algorithm *MemoryAssign*
Input: Application programs P with Register-allocated variables marked;
SRAM_Size: Size of Scratch-Pad SRAM
Output: Assignment of arrays to SRAM/DRAM

```
    AvSpace = SRAM_Size
    -- AvSpace is the free SRAM space currently available
    Let U = {array  u|u  is an array in P}
    -- U is the set of all behavioral arrays in program P
    Let W = φ   -- W is the set of arrays assigned to DRAM
    for all variables v
        if v is a scalar variable or constant
            Assign v to SRAM
            AvSpace = AvSpace − size(v)
        else
            if size(v) > SRAM_Size
                W = W ∪ {v}   -- Assign v to DRAM
    end for
    Generate compatibility graph G based on life-times of remaining arrays
    U = U − W   -- U is the set of all arrays smaller than SRAM size
    while (U ≠ φ)
        for each array u ∈ U
            Find largest clique c(u) in G such that u ∈ c(u) and
            size(v) ≤ size(u) ∀v ∈ c(u)
            Compute access density AD(u) = (Σ_{v∈c(u)} TCF(v)) / size(u)
        end for
        Assign clique c(i) to SRAM, where AD(i) = max {AD(u)|u ∈ U}
        -- Assign cluster with highest access density to SRAM
        AvSpace = AvSpace − size(c)
        -- size(c) is the size of largest array in cluster c
        X = {v ∈ U|size(v) > AvSpace}
        -- X is the set of arrays in U larger than the available space AvSpace
        W = W ∪ X -- Arrays in X are mapped to DRAM
        U = U − {v|(v ∈ c)} − X
        -- Remove from U arrays assigned to SRAM and arrays in X
    end while
    Assign arrays in W to DRAM
end Algorithm
```

Figure 5.7. Memory Assignment of variables into SRAM/DRAM

access density $AD(u)$ (i.e., largest $\sum TCF$, since maximum memory space required = size(u)). This problem is easily seen to be NP-Hard by using an instance where $TCF(u) = 1$ and size(u) = 1 for every node u, and inferring a reduction from the *Maximal Clique Problem* [GJ79]. We use a greedy heuristic [TS86], which starts with a single node clique consisting of node u, and iteratively adds neighbour v with maximum $TCF(v)$, provided size(v) $\leq$ size(u) and v has an edge with all nodes in the clique constructed so far.

After generating the cliques for each array u, we assign the one with the highest access density to the SRAM, following which, we delete all nodes in the clique and connecting edges from the graph G. In case of more than one array with the same access density, we choose the array with the larger *TCF* value. After every SRAM assignment, we assign all the unassigned arrays that are larger than the available SRAM space, to the DRAM. We then iterate through the process until all nodes in G have been considered. Note that the clustering step is re-computed in every iteration because the overlapping nature of the clusters might cause the optimal clustering to change after a clique is removed.

Analyzing algorithm *MemoryAssign* for computational complexity, we note that determining the largest clique is the computationally dominant step, with our greedy heuristic requiring $O(n^2)$ time, where n is the number of arrays in program P. Determining the cliques for all arrays requires $O(n^3)$ time in each iteration. Since the algorithm could iterate a maximum of $O(n)$ times, the overall complexity is $O(n^4)$.

When applying algorithm *MemoryAssign* to practical applications, we notice that the number of arrays, n, is not necessarily small, but the number of arrays in a program with non-intersecting life-times is usually small. Consequently, the compatibility graph of arrays for a program tends to be sparse. This is in contrast to the corresponding graph for scalars, which is usually dense. This indicates that an exhaustive-search algorithm for determining the cliques might be acceptable. In our implementation, we use exhaustive search if the number of edges in a compatibility graph is $\leq 2n$, where n is the number of nodes, otherwise we use our greedy heuristic.

5.4 CONTEXT SWITCHING

Although an embedded system based on a processor core is likely to execute only a single program most of the time, some applications might require a context switch among different programs. In the architecture of Figure 5.2, if the entire SRAM is allocated to one program, the rest of the programs would not benefit from the presence of the SRAM in the architecture. One logical solution is to partition the SRAM space and allocate different partitions to different programs (Figure 5.8). This partitioning of the SRAM space is acceptable, because the number of concurrent programs involved is usually small in an embedded system which executes only a single application. Note that, unlike the register file, it is usually prohibitively expensive to save the contents of the entire Scratch-Pad SRAM in external DRAM on a context switch, because it would add a considerable delay to the context switching time.

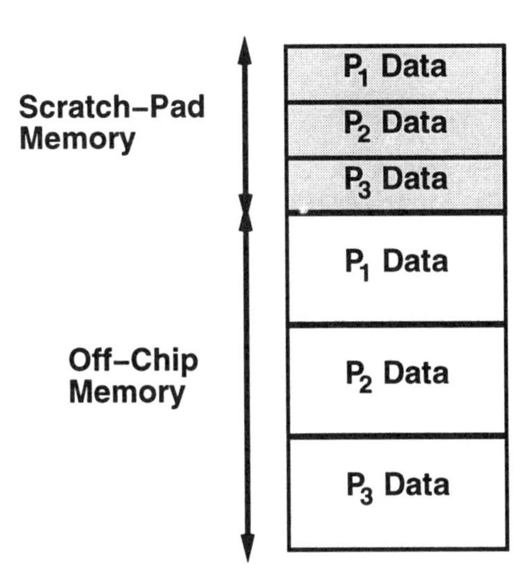

Figure 5.8. Memory space partitioned between data from different programs

To allocate the SRAM space efficiently among arrays of different programs, we use the same *TCF* metric identified in Equation 5.3. The partitioning algorithm, called *SRAMPartition*, is similar to algorithm *MemoryAssign* (Section 5.3.3). We now include arrays of all programs in the memory assignment phase. Since we have no prior knowledge of the order in which the different programs will be executed, we assume that arrays from different programs cannot share SRAM space. The SRAM space needs to be partitioned among variables in different programs only if the programs are expected to be executed in a threaded fashion. If the execution order of the programs is known to be sequential, then the entire SRAM is available to every program, and algorithm *MemoryAssign* can be used for memory assignment by considering each program separately.

5.4.1 Programs with Priorities

Some embedded systems are characterized by programs with varying priorities, which might arise from frequency of execution or the relative criticality of the programs. In such a case, it is necessary to modify algorithm *SRAMPartition* to take the relative priorities of programs $P_1 \ldots P_g$ into account. The approach we adopt in this case is to weight the TCF values for each array in a program by the priority of the respective program. That is, if the priorities of programs $P_1 \ldots P_g$ are given by $r_1 \ldots r_g$, we update $TCF(u) = TCF(u) \times r(i)$ for every array u in program P_i. The rest of the partitioning procedure remains the same as before.

5.5 GENERALIZED MEMORY HIERARCHY

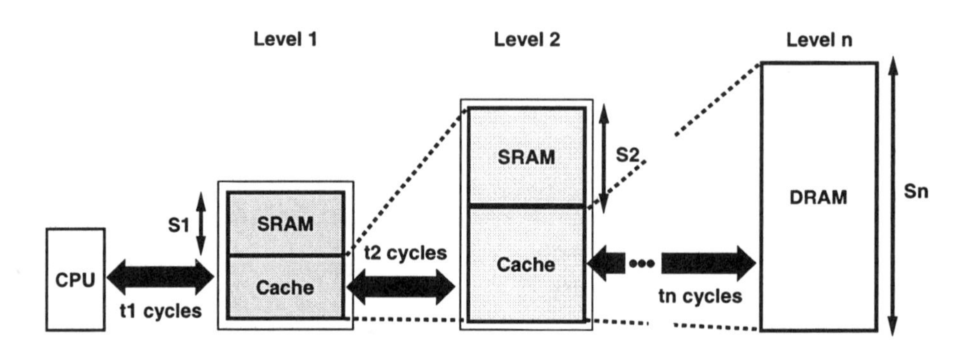

Figure 5.9. Generalized Memory Hierarchy

The architecture shown in Figure 5.2 can be generalized into an n-level hierarchy shown in Figure 5.9, with the data cache at each level acting as the interface between the levels above and below it. The total Scratch-Pad memory at level i is S_i and the time required for the processor to access data at that level is t_i cycles. Such an architecture becomes realistic with the advent of embedded DRAM technology, where large amounts of memory can now be integrated on-chip.

The memory assignment algorithms *MemoryAssign* and *SRAMPartition* can be generalized in a straightforward manner to effect memory assignment to program data in the hierarchical memory architecture shown in Figure 5.9. Since the memory access times of the different levels follow the ordering: $t_1 < t_2 < \ldots < t_n$, we store the most critical data in level 1, followed by level 2, etc. Thus, in the first iteration of the memory data partitioning, we invoke algorithm *MemoryAssign* with the SRAM size $= S_1$ and off-chip memory size $= S_2 + S_3 + \ldots S_n$. After the assignment to the first level Scratch-Pad memory, we remove the variables assigned to the first level, and continue the assignment for the remaining variables, now with SRAM size $= S_2$ and off-chip memory size $= S_3 + S_4 + \ldots S_n$. In other words, at every level i, the most critical data, as determined by the metric in Section 5.3.3, is mapped into the Scratch-Pad memory of size S_i.

The above greedy strategy is suitable when there is a relatively large number of arrays. However, an exhaustive search strategy could also be adopted for the memory assignment if the number of arrays involved is small.

5.6 EXPERIMENTS

In this section, we describe simulation experiments on several benchmark examples that frequently occur as code kernels in embedded applications, to evaluate the efficacy of our SRAM/DRAM data partitioning algorithm. We use an example Scratch-Pad SRAM and a *direct-mapped, write-back* data cache size of 1 KByte each. In order to demonstrate the soundness of our technique, we compare the performance, measured in total number of processor cycles required to access the data during execution of the example, of the following architecture and algorithm configurations:

(A) Data cache of size 2K: in this case, there is no SRAM in the architecture.

(B) Scratch-Pad memory of size 2K: in this case, there is no data cache in the architecture, and we use a simple algorithm that maps all scalars, and as many arrays as will fit into the SRAM, and the rest to the off-chip memory.

(C) Random Partitioning: in this case, we used a 1K SRAM and 1K Data cache, and a random data partitioning technique (variables were considered in the order they appeared in the code, and mapped into SRAM if there was sufficient space).

(D) Our Technique: here we used a 1K SRAM and 1K data cache, and algorithm *MemoryAssign* for data partitioning.

The size 2K is chosen in **A** and **B**, because the area occupied by the SRAM/cache would be roughly the same as that occupied by 1K SRAM + 1K cache, to a degree of approximation, ignoring the control circuitry. We use a direct-mapped data cache with line size = 4 words, and the following access times:

- Time to access one word from Scratch-Pad SRAM = 1 cycle.
- Time to access one word from off-chip memory (when there is no cache) = 10 cycles.
- Time to access a word from data cache, on cache hit = 1 cycle.
- Time to access a block of memory from off-chip DRAM into cache = 10 cycles for initialization $+ 1 \times$ Cache Line Size $= 10 + 1 \times 4 = 14$ cycles. This is the time required to access the first word + time to access the remaining (contiguous) words. This is a popular model for cache/off-chip memory traffic[PH94].

5.6.1 Benchmark Examples

Table 5.1 shows a list of benchmark examples on which we performed our experiments, and their characteristics. Columns 1 and 2 indicate the name and a brief description of the benchmarks. Columns 3 and 4 give the number of scalar and variables respectively, in the behavioral specifications. Column 5 gives the total size of the data in the benchmarks.

Table 5.1. Profile of Benchmark Examples: Scratch-Pad Memory

Benchmark	Description	No. of Scalars	No. of Arrays	Data Size (Bytes)
Beamformer	Radar Application	7	7	19676
Dequant	Dequantization Routine (MPEG)	7	5	2332
FFT	Fast Fourier Transform	20	4	4176
IDCT	Inverse Discrete Cosine Transform	20	3	1616
MatrixMult	Matrix Multiplication	5	3	3092
SOR	Successive Over-Relaxation	4	7	7184
DHRC	Differential Heat Release Computation	28	4	3856

Beamformer [PD95a], a DSP application, represents an operation involving temporal alignment and summation of digitized signals from an N-element antenna array. *Dequant* is the de-quantization routine in the MPEG decoder application [Gal91]. *IDCT* is the Inverse Discrete Cosine Transform, also used in the MPEG decoder. *SOR* is the Successive Over-Relaxation algorithm. *MatrixMult* is the matrix multiplication operation, optimized for maximizing spatial and temporal locality by reordering the loops. *FFT* is the Fast Fourier Transform application. *DHRC* encodes the Differential Heat Release Computation algorithm, which models the heat release within a combustion engine [CS93].

5.6.2 Detailed Example: Beamformer

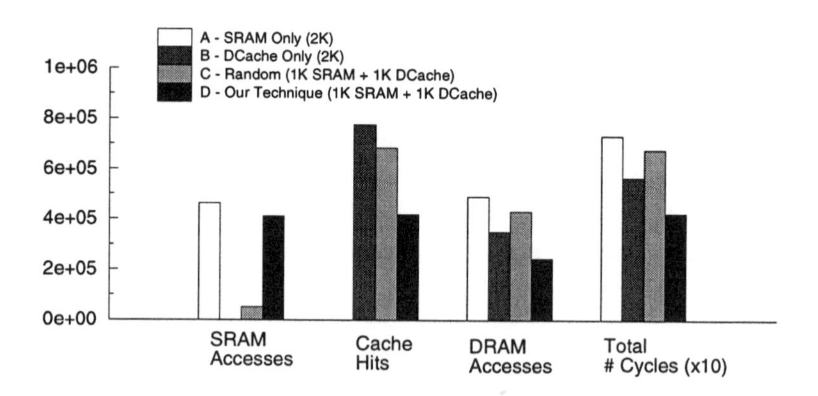

Figure 5.10. Memory Access Details for *Beamformer* Example

Figure 5.10 shows the details of the memory accesses for the *Beamformer* benchmark example. We note that configuration **A** has the largest number of *SRAM Accesses*, because the large SRAM (2K) allows more variables to be mapped into the Scratch-Pad memory. Configuration **B** has zero SRAM accesses, since there is no SRAM in that configuration. Also, our technique (**D**) results in far more SRAM accesses than the random partitioning technique, because the random technique disregards the behavior when it assigns precious SRAM space. Similarly, *Cache Hits* are the highest for **B**, and zero for **A**. Our technique results in fewer cache hits than **C**, because many memory elements accessed through the cache in **C**, map into the SRAM in our technique. Configuration **A** has a high *DRAM Access* count because the absence of the cache causes every memory access not mapping into the SRAM, to result in an expensive DRAM access. As a result, we observe that the total number of processor cycles required to access all the data is highest for **A**. Configuration **D** results in the fastest access time, due to the judicious mapping of the most frequently accessed, and conflict-prone elements into Scratch-Pad memory [1].

5.6.3 Performance of SRAM-based memory configuration

Figure 5.11 presents a comparison of the performance for the four configurations **A, B, C**, and **D** mentioned earlier, on code kernels extracted from seven benchmark embedded applications. The characteristics of the applications are described in [PDN96]. The number of cycles for each application is normalized to 100. In the *Dequant* example, **A** slightly outperforms **D**, because all the data used in the application fits into the 2K SRAM used in **A**, so that an off-chip access is never necessary, resulting in the fastest possible performance. However, the data size is bigger than the 1K SRAM used in **D**, where the compulsory cache misses cause a slight degradation of performance. The results of *FFT* and *MatrixMult*, both highly computation-intensive applications, show that **A** is an inferior configuration for highly compute-oriented applications amenable to exploitation of locality of reference. Cache conflicts degrade performance of **B** and **C** in *SOR* and *DHRC*, causing worse performance than **A** (where there is no cache), and **D** (where conflicts are minimized by algorithm *AssignMemoryType*). Our technique resulted in an average improvement of 31.4% over **A**, 30.0% over **B**, and 33.1% over **C**.

In summary, our experiments on code kernels from typical embedded system applications show the usefulness of on-chip Scratch-Pad memory in addition to a data cache, as well the effectiveness of our data partitioning strategy.

[1] Figure 5.10 shows the total number of cycles scaled down by a factor of 10

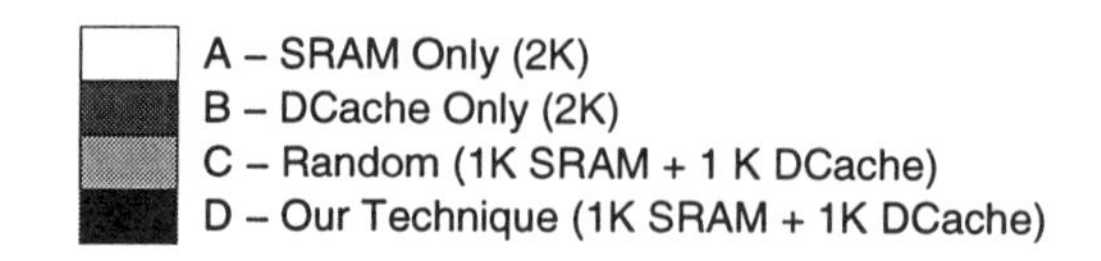

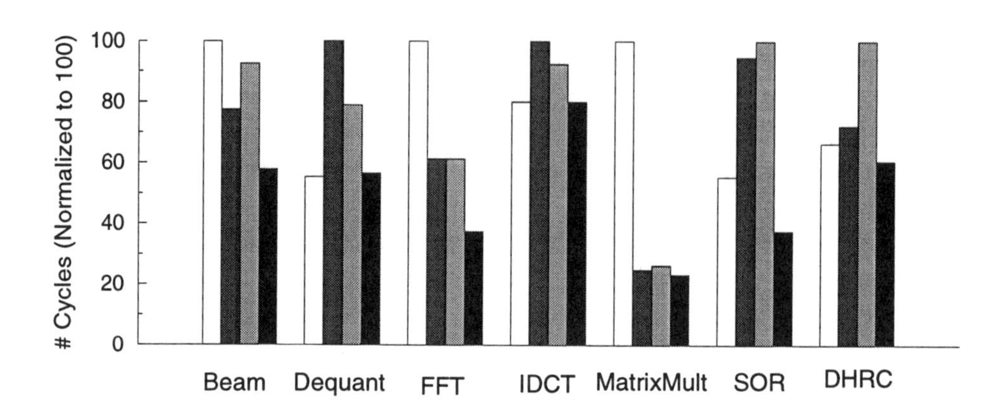

Figure 5.11. Performance Comparison of Configurations A, B, C, and D

5.6.4 Performance in the presence of context switching

To study the effectiveness of our technique for handling context switching between multiple programs described in Section 5.4, we simulated the effect of context switching between three benchmark programs: *Beamformer, Dequant*, and *DHRC*, with the same SRAM/Cache configuration as in Section 5.6.3 (1KB SRAM and 1KB data cache).

In our experiment, we compare the performance of two different SRAM assignment techniques: (1) the technique outlined in Section 5.4; and (2) a random assignment technique (*Random*), where one application (out of three) is selected at random, and our technique for SRAM assignment for a single program (configuration **D** in Figure 5.11) is used to assign SRAM space. In case SRAM space is unused, another program is selected at random, and so on.

We use the following relative priorities (Section 5.4.1) for the three examples: *Beamformer* – 1; *Dequant* – 2; and *DHRC* – 3. In the experiment, the priority value was reflected in the frequency of execution of the programs. Thus, *Beamformer* was executed once, *Dequant* twice, and *DHRC* thrice.

After performing the memory assignment, we generate a trace of memory locations accessed by each program. The trace data structure is a list of nodes where each node contains the memory address, and the memory access type (READ or WRITE) for each memory access. For the three programs, *Beamformer, Dequant*, and *DHRC*, we generate 3 traces of memory accesses: T_1, T_2, and T_3 respectively. Since *Dequant* is executed twice (priority = 2), we generate a new trace T_2' for *Dequant*, by duplicating and concatenating two copies of trace T_2. Similarly, we generate T_3' for *DHRC* by concatenating three copies of the T_3 trace. We now simulate the SRAM/Data Cache activity by randomly alternating execution between the three traces: T_1, T_2', and T_3'.

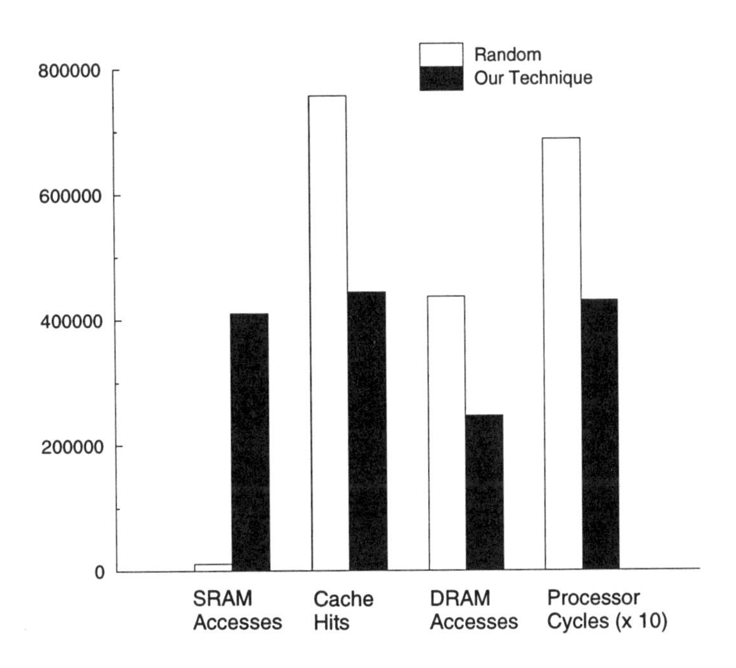

Figure 5.12. Performance in the presence of context switching between *Beamformer*, *Dequant*, and *DHRC* examples

Figure 5.12 shows the memory access details for the above experiment. The number of *SRAM accesses* is very low for *Random*, because the number of array accesses in the program whose arrays were assigned to the SRAM by *Random*, was relatively lower. *Cache Hits* were higher for *Random* because it assigned more accesses to the DRAM data. In our technique, the number of accesses to external DRAM was significantly lower because of efficient SRAM utilization. Overall, there is a 37% reduction in the total processor cycles due the memory accesses. This experiment indicates the usefulness of a judicious SRAM partitioning/assignment strategy allowing for context switching between different applications.

5.7 SUMMARY

In order to effectively use on-chip memory in embedded systems, we need to leverage the advantages of both data cache (simple addressing) and on-chip Scratch-Pad SRAM (guaranteed low access time) by including both types of memory modules in the same chip. In this chapter, we presented a strategy for partitioning scalar and array variables in embedded code into Scratch-Pad SRAM and data cache, that attempts to minimize off-chip memory traffic. We also presented extensions of the technique to handle context switching between different application programs, as well as a generalized hierarchical memory architecture.

6 MEMORY ARCHITECTURE EXPLORATION

6.1 MOTIVATION

An important feature of embedded processor-based design for systems-on-chip is that the CPU core is decoupled from the on-chip memory, and it is the system designer's responsibility to instantiate an appropriate on-chip memory configuration in his design. Traditionally, cache configuration for commercial microprocessors has been determined by conducting experiments on benchmark examples. Since general-purpose microprocessors in a PC or workstation environment have to execute a large variety of application software, a usual and effective means to improve performance has been to increase the cache size to the extent allowable by silicon area constraints. This strategy is acceptable because, in the absence of advance knowledge of the application programs that will execute on the processor, it is fair to assume that an increase in cache size will, in most cases, lead to improvement in performance, because more instruction and data can be held in local memory using larger caches, leading to possibly greater degree of reuse. However, in the embedded processor domain, there is often only a single application, and it is possible to conduct a more thorough analysis of the application to determine the best memory configuration. When coupled with an aggressive compiler that exploits the knowledge of this architecture, the impact on the overall design is significant. For instance, if an analysis of the application reveals that the data cache hit ratio is not likely to improve for cache sizes larger than 1 KByte, the information can be utilized to allocate expensive on-chip silicon area to other hardware resources, instead of an unnecessarily large cache.

An analytical prediction of the expected performance for various memory architectures has several advantages over a simulation-based approach. First, it is orders of magnitude faster and is independent of the data size, unlike the simulation. Moreover, such an analysis is extremely useful in the initial stages of system design, when decisions on hardware/software partitioning are made. Although embedded systems afford longer program compilation times, the initial exploration phase of hardware/software trade-offs is, nevertheless, time consuming. The faster prediction made possible by an analytical technique allows a much larger design space to be explored, as opposed to a simulation-based technique.

In this chapter, we discuss an exploration strategy for determining an efficient on-chip data memory architecture – characterized by Scratch-Pad memory size and cache parameters – based on an analysis of a given application, so as to reduce off-chip memory traffic. The components in the generic embedded system architecture addressed in the exploration are the same as those highlighted in Figure 5.1.

6.2 ILLUSTRATIVE EXAMPLE

We illustrate the strategy for architectural exploration and optimization of on-chip memory on CONV, a convolution program frequently used in Image Processing tasks such as edge detection, regularization, and morphological operations [BMMZ95]. The code for CONV is shown in Figure 6.1.

A small (4×4) matrix of coefficients, *mask*, slides over the input image, *source*, covering a different 4×4 region in each iteration of y, as shown in Figure 6.2. In each iteration, the coefficients of *mask* are combined with the region of the image currently covered, to obtain a weighted average, and the result, *acc*, is assigned to the pixel of the output array, *dest*, in the center of the covered region.

We illustrate a typical memory exploration situation through the following problem: given a maximum amount, say 4 KB of on-chip memory space, find an efficient utilization of the space, in terms of data cache size, cache line size, and Scratch-Pad memory size, with the objective of minimizing the number of processor cycles required to access the arrays from memory. We assume that scalar variables are assigned to registers; the data cache is direct-mapped and write-through [PH94]; and the largest allowed cache line size is 128 Bytes.

Since we use a write-through cache, in which memory writes do not interfere with the cache contents in case of cache misses, the access to *dest* does not cause cache conflicts. However, if the two arrays *source* and *mask* were to be accessed through the data cache, the performance would be affected by cache conflicts. Further, an

```
# define N 128
# define M 4
# define NORM 16
int source[N][N], dest [N][N];
int mask [M][M];
int acc, i, j, x, y;
:
for (x = 0; x < N - M; x++)
    for (y = 0; y < N - M; y++) {
        acc = 0;
        for (i = 0; i < M; i++)
            for (j = 0; j < M; j++)
                acc = acc + source[x+i][y+j] * mask[i][j];
        dest[x+M/2][y+M/2] = acc/NORM;
    }
```

Figure 6.1. Procedure CONV

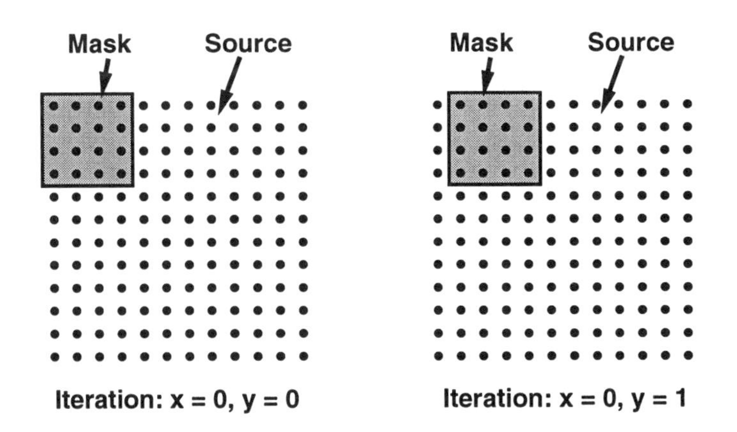

Figure 6.2. Memory access pattern in CONV example

associative cache, by itself, will not eliminate the problem in general, because most practical caches have a limited associativity, and the general situation might require an associativity as large as the number of conflicting arrays.

The conflict problem can be solved by storing the small *mask* array in the Scratch-Pad memory. This assignment eliminates all conflicts in the data cache – the data cache is now used for memory accesses to *source*, which are very regular. Also, since $M(= 4)$ rows of the *source* array are *active* at any point in time, the data cache need be only as large as $M(=4)$ rows of the source array. As mentioned before, the *dest* array is not allocated any storage in a write-through cache. If a *write-back* cache were used instead, we would allocate an equal cache space to *dest*, along with an appropriate memory assignment so that *source* and *dest* do not conflict in the cache [PDN97b]. Thus, we select a data cache of size $M \times N = 4 \times 128 = 512$ words = 2 KB. Since the accesses to *source* have good spatial locality, we select the largest allowed cache line, i.e., 128 bytes. The Scratch-Pad memory size is 4×4 words = 64 Bytes.

6.3 EXPLORATION STRATEGY

In our formulation, a local (on-chip) memory architecture for an application is defined as a combination of:

- The total size of on-chip memory used for data storage.
- The partitioning of this on-chip memory into:
  - Scratch-Pad SRAM, characterized by its size; and
  - Data cache, characterized by
    - the cache size; and
    - the cache line size

The basic algorithm for memory architecture exploration is summarized in Figure 6.3.

Algorithm *MemExplore*
    L_1: **for** on-chip memory size T (in powers of 2)
        L_2: **for** cache size C (in powers of 2, $< T$)
            SRAM Size $S = T - C$
            DataPartition(S)
            L_3: **for** line size L (in powers of 2, $< C$, $<$ *MaxLine*)
                Estimate Memory Performance
            end for
        end for
        Select (C, L) that maximizes performance
    end for
end Algorithm

Figure 6.3. Overall algorithm for memory architecture exploration

For each candidate on-chip memory size T (loop L_1), we consider different divisions of T (loop L_2) into cache (size C) and Scratch-Pad SRAM (size $S = T - C$), selecting only powers of 2 for C. Procedure *DataPartition* essentially invokes algorithm *MemoryAssign* (Section 5.3.3) for partitioning program variables into Scratch-Pad memory and cache. In the next section, we describe the memory performance estimation step. For each T, we select the (C, L) pair that is estimated to maximize performance. Finally, we perform the memory address assignment of the variables using the cache and SRAM parameters selected using the algorithms in Sections 4.2.2, 4.2.3 and 5.3.3.

6.4 MEMORY PERFORMANCE ESTIMATION

The heart of the memory exploration algorithm is the performance estimation technique that is parameterized by the cache line size, cache size, Scratch-Pad SRAM size, and total on-chip memory size. There is a trade-off in sizing the cache line. If the memory accesses are very regular and consecutive, i.e, exhibit good spatial locality, a longer cache line is desirable, since it minimizes the number of off-chip accesses and exploits the locality by pre-fetching elements that will be needed in the immediate future. On the other hand, if the memory accesses are irregular, or have large strides, a shorter cache line is desirable, as this reduces off-chip memory traffic by not bringing unnecessary data into the cache. The maximum size of a cache line is the DRAM page size, which is usually a power of two. Thus, for example, for a DRAM with line size 1 KB = 256 words, there are a maximum of 9 (2^0 through 2^8) alternatives for the cache line size.

Suppose there are N scalar variables stored in off-chip memory, accessed M times in the program. We store all scalar variables in consecutive locations in memory. Since accesses to scalars invariably constitute a small fraction of the total accesses, we make the simplifying assumption that there is only one cache miss (a compulsory miss that occurs when the variable is first accessed into the cache) for every cache line containing scalars. Although this assumption looks too optimistic, it is actually reasonable when combined with our partitioning strategy for scalar and array variables – most of the scalars get mapped to the register file and Scratch-Pad memory, and not to the cache. Since there are N scalars, we require $\lceil \frac{N}{L} \rceil$ cache lines for them for a cache line size of L, i.e., there are $\lceil \frac{N}{L} \rceil$ cache misses, and consequently, $M - \lceil \frac{N}{L} \rceil$ cache hits for the M accesses. A memory access resulting in a cache hit requires one cycle, while a cache miss entails a delay of $(K + L)$ processor cycles, where K is a constant (usually 10–20 in modern processors [PH94]). Thus, the total number of processor cycles required for the M accesses to scalar variables is:

$$\text{Cycles (Scalars)} = (K + L)\left\lceil \frac{N}{L} \right\rceil + M - \left\lceil \frac{N}{L} \right\rceil \qquad (6.1)$$

We determine an estimate of the processor cycles required to access the array elements by first dividing the application program into loop nests. Straight line code

for $i = 1$ **to** $M - 1$ **step** 1
  for $j = 1$ **to** $M - 1$ **step** 1
    $A[i][j] = A[i][j] + A[i-1][j] + A[i+1][j] +$
      $A[i][j-1] + A[i][j+1] + B[i] + C[j][i]$

(a)

L_1 : **for** $i_1 = l_1$ **to** h_1 **step** s_1
  L_2 : **for** $i_2 = l_2$ **to** h_2 **step** s_2
    ...
    L_m : **for** $i_m = l_m$ **to** h_m **step** s_m
      Read $a[i_1][i_2] \ldots [i_m]$
      ...
      L_n : **for** $i_n = l_n$ **to** h_n **step** s_n
        Read $b[i_1 + k_1][i_2 + k_2] \ldots [i_n + k_n]$
        Read $b[i_1 + k'_1][i_2 + k'_2] \ldots [i_n + k'_n]$

(b)

Figure 6.4. (a) Example loop (b) General n-level loop nest

is treated as a singly-nested loop with an iteration count of one. Multi-dimensional arrays are assumed to be stored in row-major format.

Consider the example loop shown in Figure 6.4(a). $B[i]$ is reused in different j-iterations, so it is moved up into the i-loop. $A[i][j], A[i][j-1]$, and $A[i][j+1]$ exhibit *group-spatial* reuse [WL91] – the cache line accessed by one reference will usually also contain the data for the others. $A[i-1][j]$ and $A[i+1][j]$ have *self-spatial* reuse [WL91] – each can reuse the cache line it accessed in the previous j-iteration. Finally, reference $C[j][i]$ has no reuse. The memory references are grouped into *reuse equivalence classes* – each class consists of memory references that exhibit self-spatial or group-spatial reuse. This classification is used in [WL91] to aid in loop transformation procedures, such as skewing, reordering, etc., by enabling a comparison of the reuse properties of different candidate transformations. We discuss a refinement of the above procedure in order to utilize the reuse analysis for our line size selection problem.

6.4.1 Refinement to self-spatial locality

In Figure 6.4(a) reference $B[i]$, after being moved to the i-loop, was assumed to have spatial reuse. However, this reuse can occur only if the cache line corresponding to

$B[i]$ is still present in the cache when the next i-iteration begins, and has not been flushed out by the intervening accesses.

We generalize the above reuse condition for the example n-level nested loop of Figure 6.4(b). We assume that the loop bounds, and branch probabilities for conditionals are statically known, as is the usual case in many embedded applications. If such information is not known, we have to rely on profiling statistics. We use the following locality criterion: spatial locality for the reference $a[i_1][i_2]\dots[i_m]$ in the level-m loop (L_m) can be exploited if the total number of memory accesses in inner loops (i.e., loops $L_{m+1}\dots L_n$) is less than the cache size. Let the number of memory references at loop level j be c_j. For example, in Figure 6.4(b), $c_m = 1$ (for the single memory read), and $c_n = 2$ (for the two memory reads). The number of iterations (r_j) of loop level L_j is given by: $r_j = \left\lceil \frac{h_j - l_j + 1}{s_j} \right\rceil$. Thus, the sufficient condition for utilizing locality for the reference at nesting level m is:

$$\sum_{i=m+1}^{n} c_i \left(\prod_{j=m+1}^{i} r_j \right) < \mathit{CacheSize} \tag{6.2}$$

Note that the condition above, which involves the *number* of elements accessed, is an approximation of the cache behavior, for it assumes a fully associative cache with a perfect replacement policy. We incorporate the effect of cache conflicts that occur in limited associativity caches in Section 6.4.3. *CacheSize* in Equation 6.2 is in words. To be more exact, the RHS of the equation should be ($\mathit{CacheSize} - L$), where L is the cache line size.

6.4.2 Refinement to group-spatial locality

In Figure 6.4(a), reference $A[i-1][j]$ and $A[i][j]$ are assumed to have no spatial or temporal locality because they are in different rows. However, this is a pessimistic assumption. If one complete row of A fits in the cache, then the data read by $A[i][j]$ in one iteration can be reused by the $A[i-1][j]$ reference in the next i-iteration.

We formalize the above observation into a general condition for predicting reuse for the two b-references in Figure 6.4(b). We first determine a feasibility condition that needs to be met if the two references are to access the same cache line in different iterations. The sharing can occur only if all the index expressions in the higher dimensions ($1\dots n-1$) match exactly for different values of the loop index, and the reuse-dimension (lowest dimension) resolves to expressions that differ in less than the cache line size. For example, we require that $(i_1 + k_1)$ in one iteration, say $i_1 = I$, be equal to $(i_1 + k_1')$ in some other iteration, say $i_1 = I + s_1 \cdot t$, for an integer t, since the two iteration numbers are separated by a multiple of the loop stride, s_1. We need to have: $I + k_1 + s_1 \cdot t = I + k_1'$, i.e., $k_1' - k_1 = s_1 \cdot t$, i.e., $(k_1 - k_1') \bmod s_1 = 0$.

Generalizing for dimensions $[1 \ldots n-1]$, we have:

$$\forall j \in [1 \ldots n-1], (k_j - k'_j) \bmod s_j = 0 \tag{6.3}$$

For the reuse dimension (lowest dimension, indexed by i_n), we do not need an exact match, but only need that the expressions differ in less than the cache line size L. Thus, for two iterations: $i_n = I$ and $i_n = I + s_n \cdot t$, we need to have:

$$(I + k_n + s_n \cdot t) - (I + k'_n) \leq L \tag{6.4}$$

i.e.,

$$k_n - k'_n - L \leq s_n \cdot t \tag{6.5}$$

To be applied as a feasibility test, this can be rephrased as:

$$\exists l \in [1 \ldots L], \text{ such that } (k_n - k'_n - l) \bmod s_n = 0 \tag{6.6}$$

Further, we need to ensure that the number of elements brought into the cache between the two accesses to b in Figure 6.4(b), is less than the cache size. For the two index expressions $[i_1 + k_1]$ and $[i_1 + k'_1]$, the number of iterations of loop L_1 that elapse between the two expressions resolving to the same value is: $\left\lfloor \frac{|k_1 - k'_1|}{s_1} \right\rfloor$. Since c_1 elements are accessed at the first loop level, the number of elements from loop L_1 brought into the cache in the $\left\lfloor \frac{|k_1 - k'_1|}{s_1} \right\rfloor$ iterations is: $c_1 \cdot \left\lfloor \frac{|k_1 - k'_1|}{s_1} \right\rfloor$. The number of elements in inner loops accessed in each iteration of loop L_1 is: $\sum_{i=2}^{n} c_i (\prod_{j=2}^{i} r_j)$ (LHS of Equation 6.2, with $m = 1$). Thus, the total number of elements (f_1) brought into the cache before $[i_1 + k_1]$ and $[i_1 + k'_1]$ resolve to the same value, is given by:

$$f_1 = c_1 \cdot \left\lfloor \frac{|k_1 - k'_1|}{s_1} \right\rfloor \cdot \sum_{i=2}^{n} c_i \left(\prod_{j=2}^{i} r_j \right) \tag{6.7}$$

Summing over all the n dimensions, we have the sufficient condition to enable group-spatial reuse as:

$$\sum_{t=1}^{n} f_t \leq \mathit{CacheSize} \tag{6.8}$$

i.e.,

$$\sum_{t=1}^{n} \left(c_t \cdot \left\lfloor \frac{|k_t - k'_t|}{s_t} \right\rfloor \cdot \sum_{i=t+1}^{n} c_i \left(\prod_{j=t+1}^{i} r_j \right) \right) \leq \mathit{CacheSize} \tag{6.9}$$

We conclude that the two b-references in Figure 6.4(b) exhibit spatial locality (i.e., fall into the same reuse equivalence class) if they satisfy the conditions 6.3, 6.6 and

6.9. As mentioned before, *CacheSize* in the above equations could be substituted by the more exact $(CacheSize - L)$. The equations can be generalized to the case where the array index expressions are, of a more general form: $[a_j i_j + k_j]$ (where a_j and k_j are constants), as is the case with array references in many multimedia applications. In this case, the volume analysis would need to be complemented by a dependence analysis of the form described in [BENP93].

6.4.3 Refinement incorporating cache conflicts

Note that Equations 6.2 and 6.9 are approximations for cache reuse, for they ignore the possibility of cache conflicts. In this section, we outline the procedure for incorporating cache conflicts into the performance estimation.

6.4.3.1 Compatibility of Access Patterns. We call accesses to two arrays a and b *compatible* if they satisfy the property $f_a - f_b = constant$ (independent of loop variables), where f represents the array index expressions. In compiler terminology, the *dependence distance* [BENP93] between the two array accesses is zero for compatible accesses, and non-zero for incompatible accesses. This property holds for a large variety of typical array access patterns. For example, accesses to $A[i]$ and $A[i+2]$ in the same loop satisfy this property. The property also holds for two array accesses of the form $A[2i]$ and $B[2i+3]$.

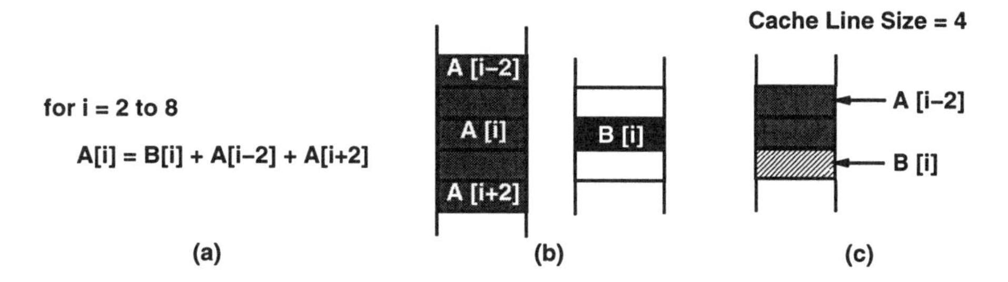

Figure 6.5. (a) Example code (b) Shaded region involved in one iteration (c) Cache mapping

If all accesses in a loop are compatible, then we can use a suitable data layout in memory to avoid cache conflicts completely (Chapter 4). Consider the code fragment in Figure 6.5(a), to be mapped into a cache with line size = 4 words. The regions of the arrays that are referenced in one iteration are shown shaded in Figure 6.5(b). A memory assignment of arrays A and B that avoids conflicts should ensure that the shaded regions of A and B never map into the same cache line. To do this, we note that the region of A involved in the array access in one iteration includes 5 words. We include the words such as $A[i-1]$ and $A[i+1]$, which are not used in the current iteration, but will be used in subsequent iterations, since they are automatically fetched into the cache line.

We note that the five consecutive words in A can occupy, in the worst case, a maximum of two lines in the cache with line size = 4 words [1]. Similarly, the one word from B can occupy one line. Thus, if we adjust the distance between $A[i-2]$ (the earliest A-word accessed in iteration i) and $B[i]$ (the earliest B-word accessed in this iteration), so that they are two lines apart when mapped into cache, we can avoid cache conflicts during loop execution. Using this strategy, when we avoid conflicts in one iteration, we also avoid conflicts in all future iterations because of the regularity (*affineness*) of array accesses (we accomplish this by selecting the maximum of cache lines that can span one iteration). In subsequent iterations, the entire pattern moves down the cache, eventually wrapping around to the top.

To facilitate our analysis of conflicts, we group the memory references in loops into *Compatibility classes*. Compatibility classes are related to the reuse equivalence classes defined earlier in that, these classes are supersets of the reuse classes. Reuse classes of more than one array with compatible access patterns are clustered into the same compatibility class.

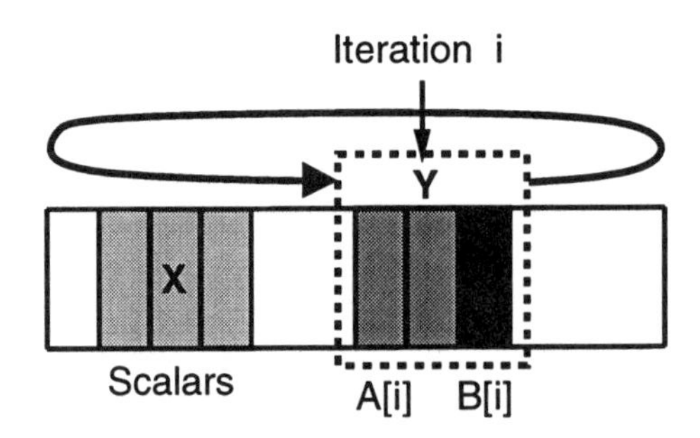

Figure 6.6. Cache conflict between scalars and arrays

6.4.3.2 Arrays Conflicting with Scalars. Scalar variables accessed in loops are typically stored in the register file, but if the number of scalars exceeds the available registers, they have to be stored in main memory, and consequently, accessed through the cache. In such a scenario, conflicts between arrays and scalars in memory are inevitable, since scalars are mapped to a fixed memory location (hence, a fixed cache location), whereas the accessed array elements map to different cache locations in different iterations. Figure 6.6 shows the cache memory where scalars in a loop map to region X of the cache, whereas elements accessed from arrays A and B map to different

[1] If the region had six words, it could span three lines. E.g., we could have $A[i]$ mapping to line j; $A[i+1]$ $\ldots A[i+4]$ mapping to line $(j+1)$; and $A[i+5]$ to line $(j+2)$.

parts of the cache in different iterations. Let M_{sc} be the number of cache lines occupied by scalars accessed in the loop, and M_i be the number of cache lines occupied by array elements accessed in one loop iteration. For the example in Figure 6.6, $M_A = 2$ cache lines; $M_B = 1$; M_{sc} (the no. of lines occupied by scalars) $= 3$. In this case, conflicts between scalars and arrays occur whenever region X (which is fixed) and Y (which is moving) intersect in the cache.

During execution of the loop, we have cache conflicts whenever an accessed array element maps to one of the M_{sc} lines occupied by the scalars. Thus, if there are n_a array accesses in the loop for cache size C lines, each of them might map to the scalar region with a probability M_{sc}/C, assuming the accessed region moves uniformly over the loop iterations. Hence the expected total number of conflict misses per access $= (M_{sc}/C)n_a$. Similarly, the probability of an array interfering with one of the n_{sc} scalar accesses is (M/C) (where M is defined as $M = \sum_i M_i$), so the expected number of scalar misses per iteration $= (M/C)n_{sc}$. The expected conflict miss ratio is, then, $((M_{sc}/C)n_a + (M/C)n_{sc})/(n_{sc} + n_a)$.

6.4.3.3 Conflicts among Arrays in Different Equivalence Classes. When multiple arrays are accessed with different access patterns in a loop (i.e., they belong to different compatibility classes), it is difficult to formalize a strategy to avoid conflicts. To compute an estimate of the miss ratio in this case, we assume a uniform distribution of the memory accesses in the cache. This assumption usually holds for data sizes that are much larger than the cache.

We first divide the arrays into compatibility classes $S_1 \ldots S_g$. Each class covers a total of M_{S_j} lines in cache, where M_{S_j} is the sum of the cache lines occupied by the constituent arrays in class S_j. The probability of the arrays interfering with any of the n_{sc} scalars in cache is M/C where $M = \sum_j M_{S_j}$. Hence, the expected number of scalar misses is: $(M/C)n_{sc}$,. Similarly, the probability that an access to any element in class S_j results in a miss, is the probability that it conflicts with any scalar, or any of the other classs. Thus, it could conflict if it were to map to any of the $M_{sc}+(M-M_{S_j})$ locations occupied by the scalars and remaining arrays. In other words, the probability of a conflict miss for class $S_j = (M_{sc} + M - M_{S_j})/C$.

If the miss ratio in loop L_j is predicted to be h_j (where h_j is the sum of miss ratios computed in Sections 6.4.3.2 and 6.4.3.3 the above computations) the expected conflict count is given by $h_j \cdot \left(\prod_{i=1}^{j-1} r_i\right)$. The estimated number of processor cycles wasted due to conflicts is given by:

$$\mathit{EstConflict}(j) = h_j \cdot \left(\prod_{i=1}^{j-1} r_i\right) \cdot (K + L) \tag{6.10}$$

where K and L are as defined in Equation 6.1.

6.4.4 Computation of Processor Cycles

Based on the analysis in the previous sections, we determine the number of processor cycles, T_{ji}, due to each reuse equivalence class R_{ji} in loop level L_j, as in the loop nest of Figure 6.4(b). Let the number of memory references in R_{ji} be $|R_{ji}|$. For the r_j iterations of loop L_j, the total number of memory accesses for class R_{ji} is: $r_j \cdot |R_{ji}|$, of which $\frac{r_j \cdot |R_{ji}|}{L}$ are cache misses, and the rest $(r_j \cdot |R_{ji}| - \frac{r_j \cdot |R_{ji}|}{L})$ are hits. Since a cache hit costs 1 cycle and a miss $(K + L)$ cycles, we have:

$$\begin{aligned} T_{ji} &= \frac{r_j \cdot |R_{ji}|}{L}(K+L) + r_j \cdot |R_{ji}| - \frac{r_j \cdot |R_{ji}|}{L} \\ &= r_j \cdot |R_{ji}| \cdot \left(\frac{K-1}{L} + 2\right) \end{aligned} \tag{6.11}$$

For the set E of all the equivalence classes at level j, we have, the total cycles, T_j given by: $T_j = \sum_i T_{ji}$.

The number of accesses with no reuse at loop level L_j is given by: $c_j - \sum_i |R_{ji}|$. The number of cycles, Q_j, spent in accessing elements with no reuse is given by:

$$Q_j = (K + L) \cdot \left(c_j - \sum_i |R_{ji}|\right) \tag{6.12}$$

Finally, the number of processor cycles, P_j, wasted due to cache conflicts for accesses in loop L_j is given by:

$$P_j = r_j \cdot \mathit{EstConflict}(j) \tag{6.13}$$

Thus, the total time spent accessing memory data at level L_j is given by: $(T_j + Q_j + P_j)$. For the entire loop nest, we multiply the total cycles at each loop level by the number of times the loop L_j is executed. Thus, the total cycles, C for the loop nest under consideration is given by:

$$C = \sum_{j=1}^{n} \left((T_j + Q_j + P_j) \cdot \prod_{i=1}^{j-1} r_j\right) \tag{6.14}$$

To determine the number of cycles required for all memory accesses in the program, we take the cumulative estimate over all loop nests. We use this estimate in the memory performance estimation step in algorithm *MemExplore* (Figure 6.3).

6.5 EXPERIMENTS

We present some exploration results on sample application routines from the image processing and digital signal processing domain. The benchmark examples are:

Histogram – a histogram evaluation routine, commonly used in image enhancement algorithms [GW87]; *Lowpass* – another image processing algorithm used for accentuating low frequencies in an image, so that the resulting image has lower changes between neighboring pixel values [PD95a]; and *Beamformer* – a radar application, involving the summation of digitized signals from an antenna array [PD95a]. We present below a comparison of the estimated memory cycles for the benchmarks with the actual memory cycles. The estimated cycles are determined by the computations in Section 6.3. The actual cycles are measured by our cache/SRAM simulator, which takes as input a stream of memory addresses generated during execution of a benchmark, and reports the number of accesses to off-chip memory. For our experiments, we use a penalty of 10 processor cycles for off-chip memory accesses (i.e., $K = 10$ in Equations 4.2.2 and 6.12).

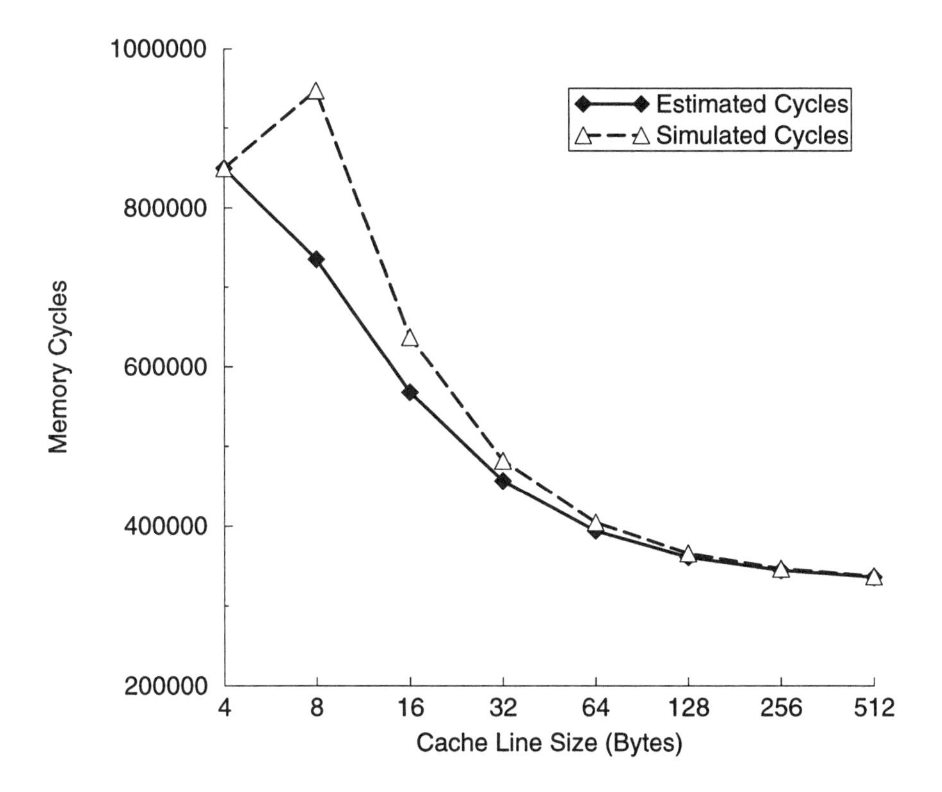

Figure 6.7. Variation of memory performance with line size for fixed cache size of 1 KB (*Histogram* Benchmark)

Our first experiment focuses on loop L_3 of the *MemExplore* exploration algorithm (Figure 6.3), where we study variation of the memory performance with the cache

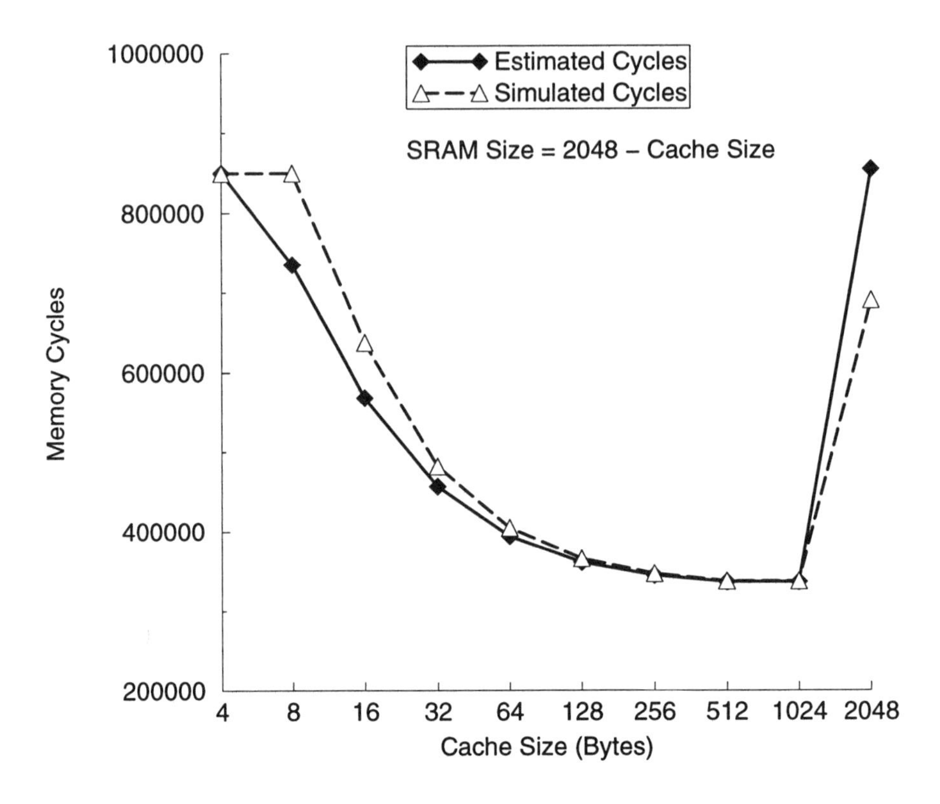

Figure 6.8. Variation of memory performance with different mixes of cache and Scratch-Pad memory, for total on-chip memory of 2 KB (*Histogram* Benchmark)

line size, for a given cache size. Figure 6.7 shows a comparison between the actual simulation result and the estimated number of processor cycles for the memory accesses in the *Histogram* routine, for a fixed cache size of 1 KB. We note that the estimated performance very closely follows the actual performance, except for one line size (8 bytes). The best cache line size selected by our algorithm (512 bytes), is also the best line size parameter determined from the simulation.

Figure 6.8 illustrates another slice of the exploration space on the same benchmark, where the estimated memory performance for different divisions of a fixed total on-chip memory space of 2 KB into data cache and Scratch-Pad memory, is compared against the simulated performance. The memory cycles plotted correspond to one iteration of the outer L_2 loop of *MemExplore*, where the best cache line size (from loop L_3) is selected for each candidate cache size. For each selected cache size, the corresponding Scratch-Pad SRAM size is given by:

$$\text{Scratch-Pad SRAM size} = 2\text{ KB} - \text{Cache Size} \tag{6.15}$$

The points on the left and right extremes represent divisions incurring severe cache conflicts (for cache size = 2048 Bytes, the SRAM size is 0, causing unavoidable conflicts in the cache). The estimation process gives the best division of the 2 KB space as: 1 KB cache + 1 KB Scratch-Pad memory – this selection is validated with the actual simulation results, as shown in Figure 6.8.

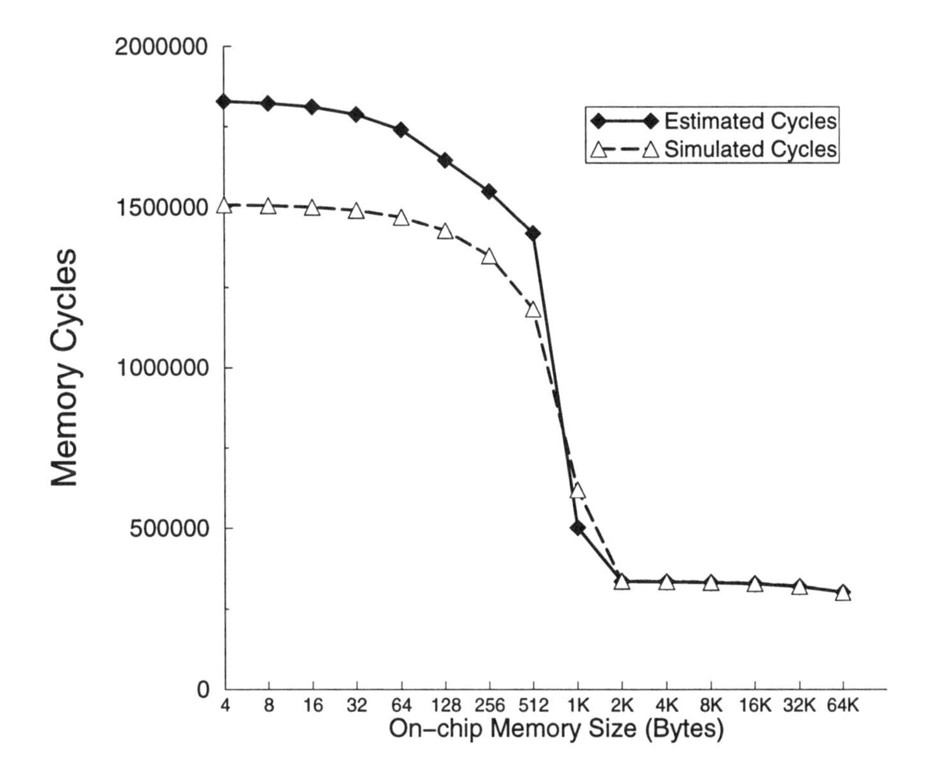

Figure 6.9. Variation of memory performance with total on-chip memory space (*Histogram* Benchmark)

Figure 6.9 shows the variation of the memory performance with the total on-chip memory space for the *Histogram* example. The y-axis shows the best performance obtained by any architecture (i.e., division into Scratch-Pad memory and cache, as well as selection of cache line size) for a given total on-chip memory space, i.e., one iteration of the outer loop L_1 in *MemExplore*. For a given application, the variation of the memory performance with the total on-chip memory is generated as feedback to the designer. The designer can then select an appropriate total on-chip memory size, based on the value beyond which no significant improvement is predicted. In the example of Figure 6.9, the total size of 2 KB is a good selection, as we observe very little improvement in cycle time beyond this cache size. Figures 6.10 and 6.11 show the results of the exploration for the *Lowpass* and *Beamformer* examples. We observe that the estimated performance curve follows the actual performance very closely. This curve can help the designer select the optimal on-chip memory size. Once the designer chooses an appropriate total memory size based on the estimation curve, the best memory architecture (consisting of the division of this space into Scratch-Pad memory and cache, as well as cache line size) is chosen automatically. Note that each point in the exploration spaces of Figures 6.9, 6.10 and 6.11 corresponds to the best architecture found for the given total on-chip memory space.

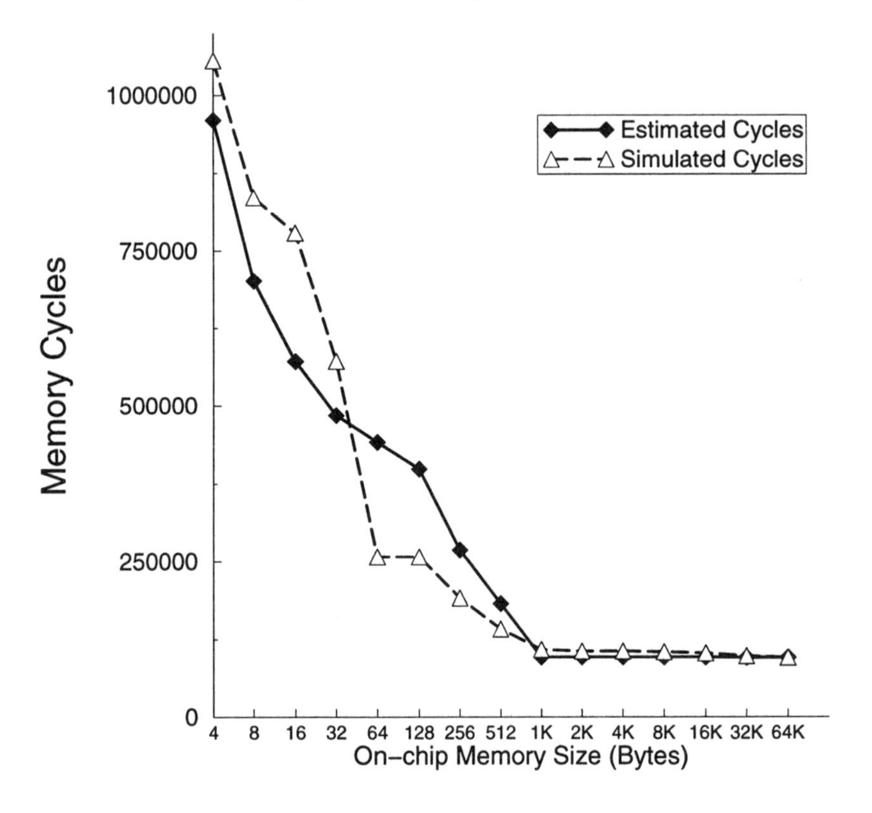

Figure 6.10. Memory exploration for *Lowpass* benchmark

The most important advantage of the analytical technique for exploring the memory performance of embedded applications is that candidate architectures can be rapidly evaluated for their memory performance. The estimation-based exploration for our experiments above required only a few seconds, which was about 1000 times faster than the simulation of the memory performance for the same set of architectures

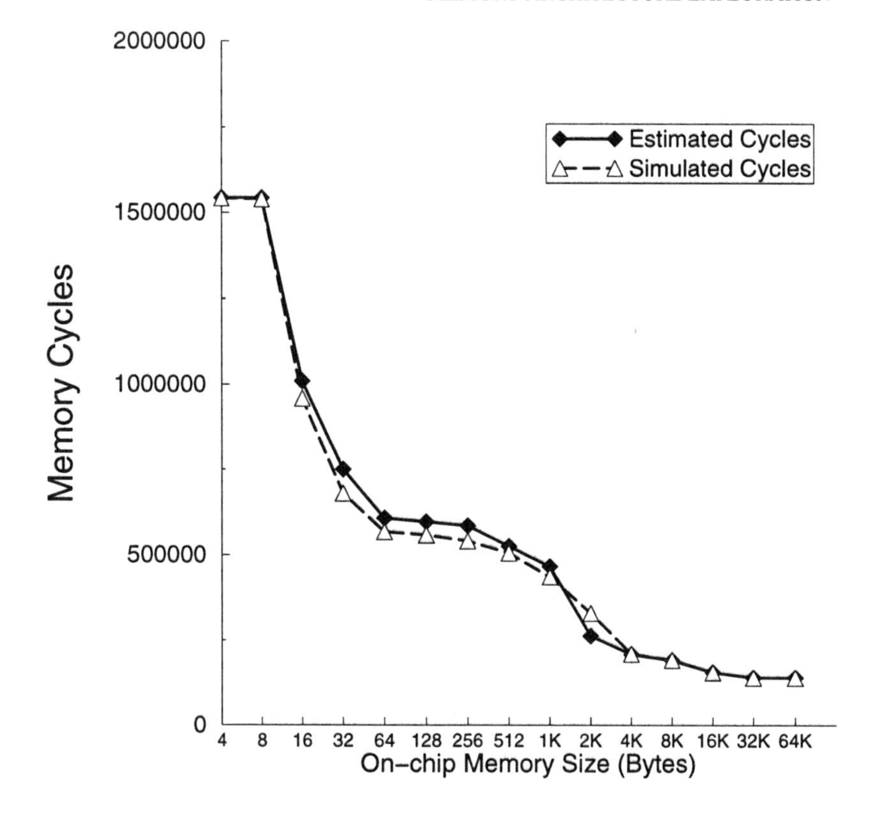

Figure 6.11. Memory exploration for *Beamformer* benchmark

explored for the data sizes we considered. For larger data sizes, the simulation times increase further, whereas the estimation time remains constant. This rapid estimation capability is very important in the initial stages of system design, where the number of possible architectures is too many, and a simulation of each architecture is prohibitively expensive.

The analytical approach discussed above can be combined with one or more final simulations to verify the predicted performance. The purpose of the analytical technique is to drastically reduce the design space so that the final design decision can be taken by the designer after a few simulations of design points around the predicted optimal result.

6.6 SUMMARY

The management of local memory space is an important problem in embedded systems involving large behavioral arrays, all of which cannot be stored on-chip. The ideal local memory architecture, consisting of a judicious division of the memory into Scratch-Pad SRAM (software-controlled) and data cache (hardware-controlled), as well as the cache line size, depends on the characteristics of the specific application. We discussed a strategy for exploration of on-chip memory architecture for embedded applications, based on an estimation of the memory performance. Thus, the exploration technique is very useful during the early stages of system design, when a large number of different possible memory architectures need to be evaluated for their performance.

7 CASE STUDY: MPEG DECODER

In Chapter 6, we discussed an exploration strategy for predicting the performance of various on-chip memory configurations for an instance of the embedded systems-on-chip architecture. We validated our exploration strategy by performing simulation experiments on small benchmark loop kernels (Section 6.5). In this chapter, we present a case study of the exploration technique on a larger application: the MPEG decoder. The MPEG is a digital video compression standard widely used in the telecommunication, computer, and consumer electronics industry today to encode and decode video images [Gal91]. We perform the memory exploration on a software implementation of the MPEG decoder. The MPEG decoder forms an interesting case study for memory exploration because it is a memory-intensive application that forms a crucial component in various graphics-related applications such as digital TV and video conferencing; its behavior is characterized by a number of interesting array access patterns, which makes it a good candidate on which to apply our memory exploration strategy. The case study demonstrates that the exploration technique is effective not just for individual loop kernels, but also more complex applications, since different loops in the MPEG decoder access the same arrays in dissimilar ways.

7.1 OVERVIEW OF MPEG DECODER

The MPEG encoding algorithm exploits spatial and temporal redundancy of consecutive video images to achieve compression. The compression algorithm uses block-based *motion compensation* for reduction of temporal redundancy and Discrete Cosine Transform (DCT)-based compression for reduction of spatial redundancy [Gal91].

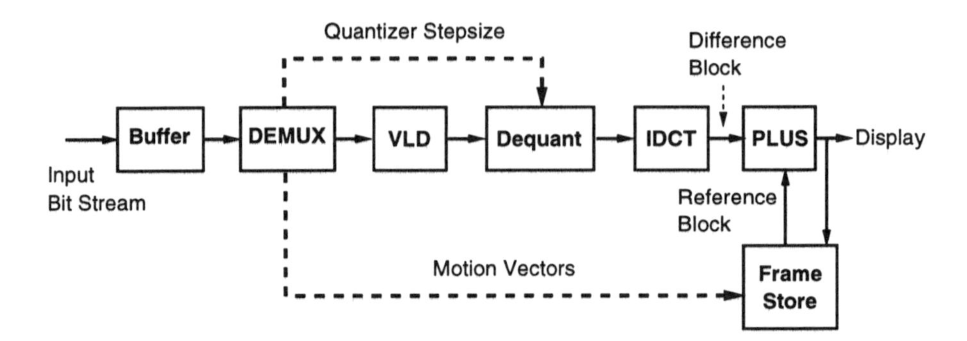

Figure 7.1. Overview of MPEG Decoder structure

Figure 7.1 shows an overview of the major functional blocks of the decoder algorithm[Tho95]. The *Buffer* stores the input bit stream so that the rest of the decoder can operate on it. The de-multiplexer (*DEMUX*) isolates the header information such as motion vectors, quantizer stepsize, and quantized DCT coefficients. The *VLD, Dequant,* and *IDCT* blocks help reconstruct the *difference-block*, which is a sort of increment to a *reference-block* stored in the *Frame Store* component. The two are then added in *PLUS* to generate the output image. This output image is stored into *Frame Store*, as well as displayed.

In the remaining sections, we first examine the memory characteristics of the individual *Dequant* and *PLUS* blocks of Figure 7.1, and finally, study the entire decoder application as a whole. We present a comparison of the estimated and simulated memory performance.

7.2 MEMORY EXPLORATION OF SAMPLE ROUTINES

7.2.1 Dequant Routine

The *Dequant* routine performs the de-quantization operation on the input quantized DCT coefficients received from the *DEMUX*[Gal91]. The code for the *Dequant* routine, which is a translation of a SpecCharts[GVNG94] model used in [Tho95], into C language, is shown in Figure 7.2. For simplicity of addressing, we convert 8-bit and 10-bit data types into the integer data type of C.

The code consists of a *for*-loop in which four arrays are accessed: *scan, curr_block, intra_quant,* and *dct_zz*. We observe that while three of the arrays, *scan, curr_block,*

```
int curr_block [64];
int scan [64];
int intra_quant [64];
int dct_zz [64];
int zigzag, zz, val;
⋮
for (i = 0; i < 64; i++) {
    zigzag = scan [i];
    zz = dct_zz [zigzag];
    val = (2 * zz * quantizer_scale * intra_quant [i]) / 16;
    if (msb(val) == 0)
        val = val - sign (val);
    if (val > 2047)
        val = 2047;
    else
        if (val < -2047)
            val = -2047;
    curr_block [i] = val;
}
```

Figure 7.2. Code for *Dequant* Routine

and *intra_quant* are accessed by the simple loop index i, the element of *dct_zz* accessed in each iteration is data dependent. The access pattern is similar to the *Histogram* example of Section 5.2. If the *scan, curr_block,* and *intra_quant* arrays are mapped into off-chip memory and accessed through the cache, they can be aligned to prevent any cache conflicts among them. However, if *dct_zz* is also accessed through the cache, data alignment does not solve the conflict problem since the accessed addresses from *dct_zz* are statically unknown. In this case, we use the technique shown in Section 6.4.3.3 to predict the memory performance. Further, the on-chip Scratch-Pad memory is useful in such a situation, because all cache conflicts can be avoided if the *dct_zz* array is stored in the Scratch-Pad memory.

7.2.1.1 Cache Line Size. Figure 7.3 shows a comparison of the estimated and actual memory performance of the *Dequant* routine, when the cache line size is varied, keeping the total cache size fixed at 1 KB.

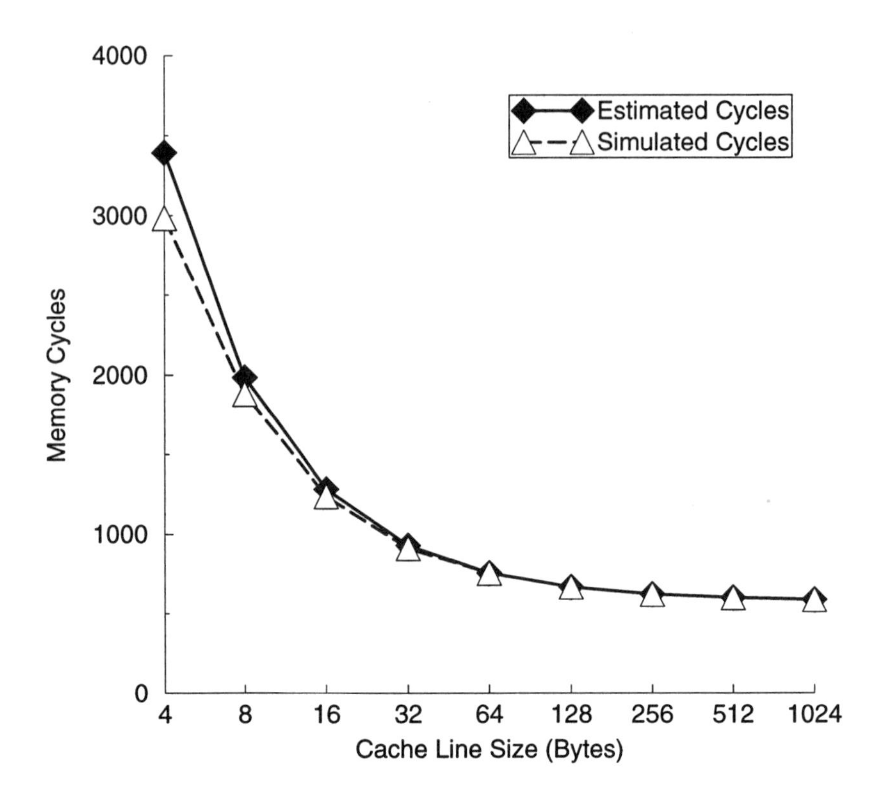

Figure 7.3. Variation of memory performance with line size for fixed cache size of 1 KB (*Dequant* Routine). No Scratch-Pad memory.

The performance improves with increasing cache line size because the combined size of all the four arrays is less than the cache size of 1 KB. In this case, there is no cache conflict, and the accesses to off-chip memory occur only due to compulsory misses when the data is first brought into the cache. Since compulsory misses are reduced by longer cache lines, we observe a monotonic performance improvement with increasing cache line size. Figure 7.3 shows that the estimated performance tracks the actual performance very well.

7.2.1.2 Cache and Scratch-Pad Memory Ratio. Figure 7.4 shows the variation of the memory performance with varying ratios of cache and Scratch-Pad memory for a fixed total on-chip memory size of 1 KB. Each point on the graph corresponds to the most favourable cache line size for the selected cache size. Thus, the points on the extreme right, corresponding to cache size = 1024 Bytes (i.e., Scratch-Pad memory size = 0) represents the best cache line size configuration for a 1024 Byte cache, i.e., the lowest (extreme right) point on the curve in Figure 7.3.

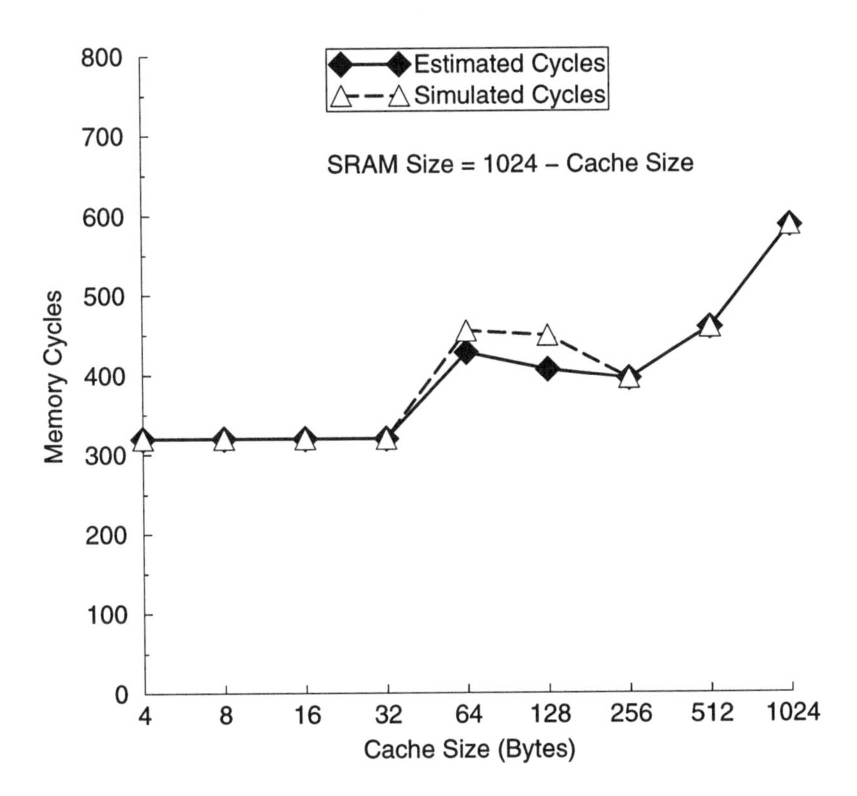

Figure 7.4. Variation of memory performance with different mixes of cache and Scratch-Pad memory, for total on-chip memory of 1 KB (*Dequant* Routine)

Since, as we observed before, the combined area of the arrays is less than 1 KB, the best performance is obtained when the arrays are all stored in on-chip Scratch-Pad memory, and the cache is not utilized at all. This is confirmed by Figure 7.4, where the best performance is predicted (and verified by simulation) for the smallest cache sizes. The performance begins to degrade when the cache size is 64 Bytes or larger. This point corresponds to the Scratch-Pad memory being too small to contain all the arrays. We observe from Figure 7.4 that the estimated memory performance is a very good indicator of the actual measured performance via simulation.

```
int curr_block [64];
int pred_block [64];
int val;
:
for (i = 0; i < 64; i++) {
    val = pred_block [i] + curr_block [i];
    if (val > 255)
        val = 255;
    else
        if (val < 0)
            val = 0;
    curr_block [i] = val;
}
```

Figure 7.5. Code for *PLUS* Routine

7.2.2 PLUS Routine

In the *PLUS* module, the difference-block (stored in array `curr_block`) is added to the reference-block (stored in array `pred_block`), to generate the output image (also stored in array `curr_block`). The code is shown in Figure 7.5.

Since the array accesses in the loop are all *compatible* (Section 6.4.3.1), we can align them in memory to eliminate cache conflicts altogether. Further, the memory performance prediction now becomes very accurate, since the array access patterns are easily analyzable.

7.2.2.1 Cache Line Size. Figure 7.6 shows the estimated and simulated memory performance as a function of cache line size, for a fixed cache size of 1 KB. We observe that the prediction is exact. The reason for the performance improvement with increasing cache line size is the same as that for the *Dequant* example: the combined sizes of the two arrays is smaller than 1 KB, hence the only misses that occur are compulsory misses, which decrease with increasing line size. Interestingly, the performance in this case worsens for the maximum cache line size (= 1024 Bytes, i.e., the cache contains only one line). This occurs because the two arrays together occupy only 512 Bytes; hence, when the line size is increased beyond 512 Bytes, unnecessary data is brought into the cache, thus increasing the memory traffic. Thus, the best cache line size for this routine is 512 Bytes.

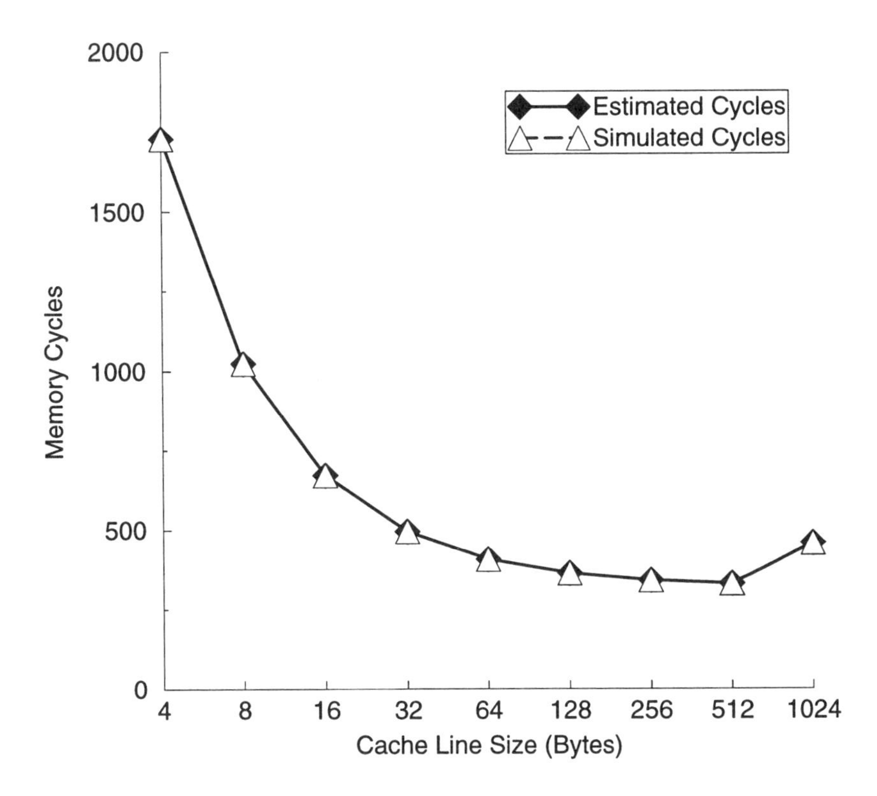

Figure 7.6. Variation of memory performance with line size for fixed cache size of 1 KB (*PLUS* Routine). No Scratch-Pad memory.

7.2.2.2 Cache and Scratch-Pad Memory Ratio. The array access pattern in Figure 7.5 also makes it possible to make accurate prediction for the performance of various configurations with varying ratios of Scratch-Pad memory and data cache. Figure 7.7 shows that the estimated memory performance is accurate for different divisions of 1 KB on-chip memory into Scratch-Pad memory and data cache. In this case, the best performance is obtained as long as there is at least 512 Bytes of Scratch-Pad memory (which is the combined size of the two arrays), i.e., as long as the cache size is $\leq (1024 - 512 = 512)$ Bytes. When the cache is 1024 Bytes, i.e., SRAM Size = 0, then the arrays have to be stored off-chip, and the performance degrades slightly because of compulsory misses.

7.3 MEMORY EXPLORATION OF MPEG DECODER

The estimated and actual memory performance of routines such as *Dequant* and *PLUS* above, will be different when they are analyzed as a part of a larger application, rather

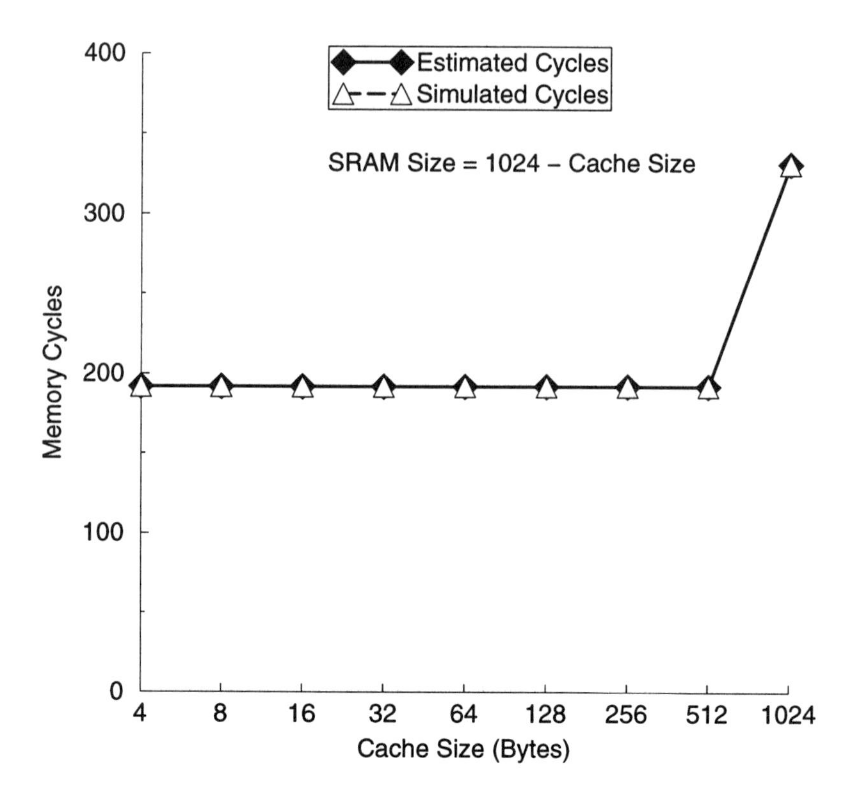

Figure 7.7. Variation of memory performance with different mixes of cache and Scratch-Pad memory, for total on-chip memory of 1 KB (*PLUS* Routine)

than individually. This is because a global analysis, in which the same array is accessed in different loops with possibly different access patterns, leads to a different memory assignment of arrays. This global analysis is necessary, since in its absence, different routines in the application will have different optimal memory configuration requirements. In this section, we study the memory performance of the entire MPEG decoder application, of which *Dequant* and *PLUS* are constituent routines. We conclude the section with a brief comparison of our approach with a simulation-based study of the MPEG-2 decoder[SL97].

We start with the SpecCharts[GVNG94] description of the MPEG decoder described in [Tho95], where each routine in Table 7.1 is modeled as a *behavior*. In the current implementation, all the routines in the MPEG decoder are flattened out for analysis; thus, the application is treated as a series of loop nests. We assume that all array sizes, loop bounds, and number of functions calls are statically known; this is true for the MPEG algorithm. Since our exploration is limited to the memory performance alone, we ignore all other computation in the loops, and focus only on the memory accesses; each array access is interpreted as a memory operation.

Table 7.1 shows some statistical information about the MPEG decoder application. In summary, there are seven routines including *Dequant* and *PLUS*, which include 12 loop nests. Column 1 gives the names of the routines; Column 2 gives the number of loop nests in each routine; Column 3 shows the number of arrays accessed in the routine; and Column 4 gives the total number of memory accesses in the routine. Of the 12 loops, four had incompatible array access patterns (Section 6.4.3.3), and the remaining 8 had compatible accesses. In some routines, such as *vld_control* and *prediction*, which involve conditional statements, we can incorporate a user-provided hint as to which path is more likely to be taken. This information could be obtained by profiling. Another possible solution is to accept probabilities of each path being taken, and incorporate them into the estimation by weighting the total estimated memory cycles by the probability of the memory access statements being executed.

Table 7.1. Characteristics of Routines in the MPEG Decoder

Name of Routine	Number of Loop Nests	Number of Arrays Accessed	Total No. of Memory Accesses
Vld_control	1	2	100
Dequant	1	4	320
Idct	4	3	2176
Prediction	3	3	512
PLUS	1	2	192
Store	1	2	128
Display	1	1	64

Since most of the scalar variables were used in local/temporary computations, we focussed our analysis on the array variables in the behavior. Table 7.2 gives details of the data involved in the decoder. Column 1 gives the array names; Column 2 gives the sizes of each array; and Column 3 shows the number of different loop nests that involve accesses to the array. Note that five of the arrays are accessed in multiple loops; for each of these arrays, the access patterns in several loops are considered when performing the memory assignment.

We ran our memory exploration algorithm of Section 6.3 on the MPEG decoder application; for each total on-chip memory size, we estimated and simulated the performance for data cache sizes ranging from one word to the total memory size. For each candidate cache, we explored cache line sizes ranging from one word to 1 KB; in general, the cache line size is limited by the page size of the off-chip DRAM. To perform our simulations, we first executed our memory address assignment algorithms

Table 7.2. Array Accesses in the MPEG Decoder

Array Name	Array Size (Bytes)	Number of Loops
curr_block	256	5
temp_block	256	2
pred_block	256	4
dct_zz	256	1
scan	256	1
intra_quant	256	1
idct_coeff	256	2
storage	12288	3
vld_length	200	1
vld_bits	200	1

for the candidate Scratch-Pad memory and cache parameters, and used the resulting array addresses to generate a trace of the data memory addresses. We then fed this trace to the cache/SRAM simulator to obtain the simulated performance result.

We validate our memory exploration algorithm of Section 6.3 on the MPEG decoder application with the following experimental strategy: for each total on-chip memory size, we estimate and simulate the performance for data cache sizes ranging from one word to the total memory size. For each candidate cache, we explore cache line sizes ranging from one word to 1 KB; in general, the cache line size is limited by the page size of the off-chip DRAM.

7.3.1 Cache Line Size

Figure 7.8 shows the variation of the estimated and actual performance with varying cache line size for a fixed cache size of 1 KB. Although the estimates are now less accurate, the estimation tracks the relative performance for different line sizes very well. The best performance is predicted to occur for a line size of 32 Bytes, which matches the simulation result, as shown in Figure 7.8. The estimation has very high *fidelity*, which is defined as the percentage of correctly predicted comparisons between points in the design space [GVNG94]. Fidelity is a very useful measure in design space exploration; an estimation with good fidelity can be effectively used to compare different different design points and arrive at the best alternative. Note that in Figure 7.8, the slopes of the line connecting two successive points in the two curves, always has the same sign, indicating that a correct decision of the best cache line size can be made by comparing the estimations of the relative merit of two neighbouring points.

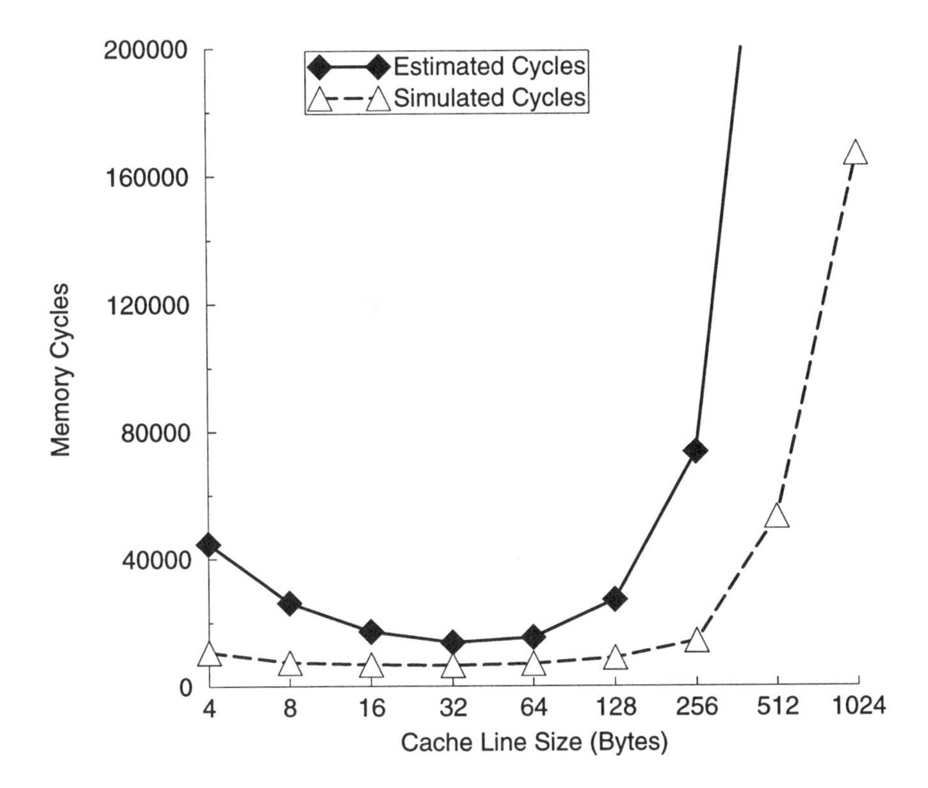

Figure 7.8. Variation of memory performance with line size for fixed cache size of 1 KB (*MPEG Decoder*). No Scratch-Pad memory.

In general, the estimated performance for the decoder is worse than the actual performance. This is because of the cache reuse occurring across different loops, which is not addressed in our estimation. For example, if an array accessed in one loop nest, is also accessed in the next loop, compulsory misses can be avoided if the array elements are still present in the cache when the second loop commences execution. However, in our estimation (Section 6.4.4), we assume that compulsory misses occur afresh in each loop. This topic requires further investigation: a heuristic that makes an intelligent guess regarding compulsory misses will help enhance the accuracy of the prediction.

7.3.2 Cache and Scratch-Pad Memory Ratio

Figure 7.9 shows the variation of the estimated and measured memory performance for different ratios of cache and Scratch-Pad memory, for a fixed total on-chip memory size of 1 KB. The estimate follows the shape of the simulation curve (i.e., has good fidelity), except for the two cache sizes: 64 Bytes and 512 Bytes. In the estimation, 64 Bytes is predicted to perform better than 32 Bytes, whereas in the simulation, 32 Bytes is better. The estimate predicts the best cache sizes as 256 Bytes, whereas simulation results indicate 512 Bytes as the best cache size (i.e., Cache : Scratch-Pad memory ratio of 1 : 1). The relatively poor estimation for 512 Bytes cache size is due to the issue of compulsory misses discussed earlier. Data reuse across different loops becomes more significant for larger caches, leading to wider variations between estimated and simulated performance.

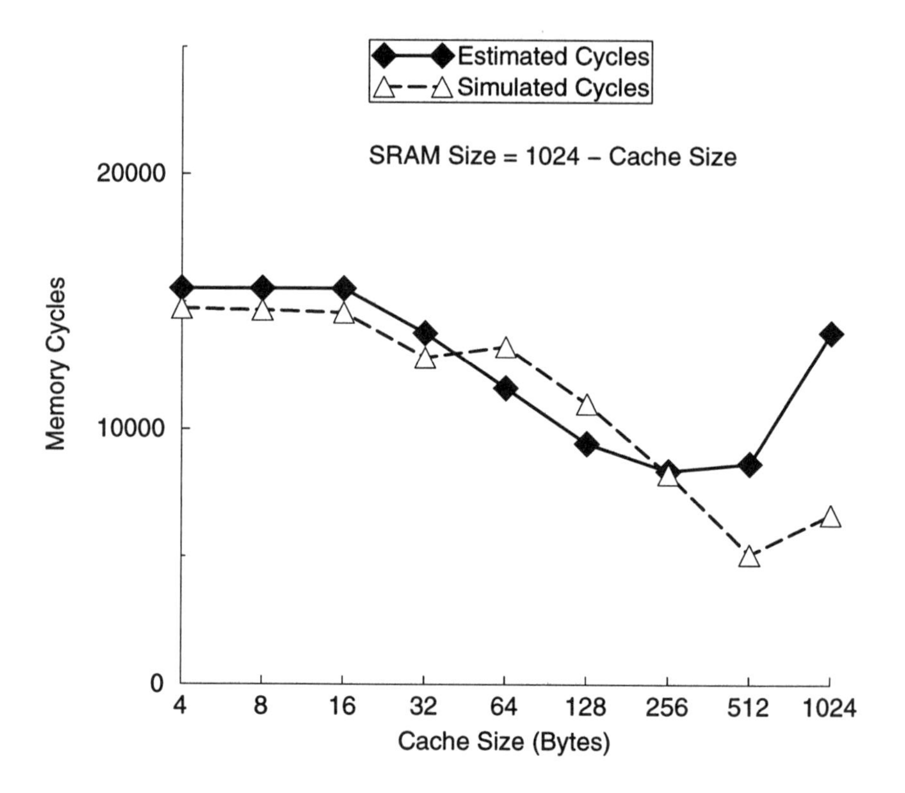

Figure 7.9. Variation of memory performance with different mixes of cache and Scratch-Pad memory, for total on-chip memory of 1 KB (*MPEG Decoder*)

Thus, as mentioned in Section 6.5, our recommendation is that the analytical exploration phase be augmented by a final simulation of the predicted best configuration, as well as a few neighbouring points on the estimation curve, especially points for which the estimated performance is very close, e.g., cache sizes 256 and 512 Bytes in Figure 7.9. Note that this is still significantly faster than an exhaustive simulation, because the latter case entails not only a full simulation of all cache sizes (and all line sizes for each cache size), but also an exponential number of combinations of array variables assigned to the Scratch-Pad memory for each configuration.

Figure 7.9 clearly demonstrates the utility of the Scratch-Pad memory in the embedded system architecture. The simulation confirms that the memory configuration corresponding to cache size = 1024 Bytes (i.e., no Scratch-Pad memory: the points on the extreme right in Figure 7.9) does not lead to the best performance. Instead, the best performance is obtained in this case when the 1024 Bytes of on-chip data memory is split into equal halves of cache and Scratch-Pad memory so that critical data can be stored in the Scratch-Pad memory.

7.3.3 Total On-chip Memory Size

Figure 7.10 shows the estimated and simulated memory performance for the decoder, for various total on-chip memory sizes varying from 4 Bytes to 64 KB. Each point on the curve corresponds to the best division of the corresponding total size into cache and Scratch-Pad memory.

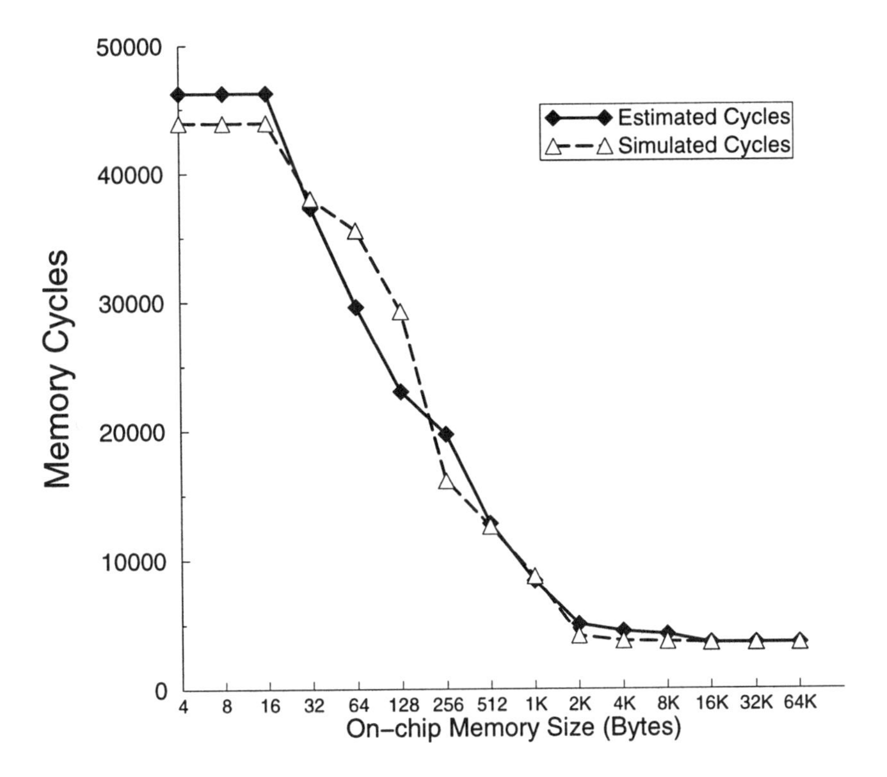

Figure 7.10. Variation of memory performance with total on-chip memory space (*MPEG Decoder*)

For very small on-chip memory sizes, the performance is expectedly poor because of consistently occurring cache conflicts. Similarly, for large memory sizes, the performance is the best because all the data fits on-chip. Note that if all the data fits on-chip, then the best configuration is to store all of it in the Scratch-Pad memory, and not use a cache at all. For the memory sizes in between, the estimated performance follows the simulation results very closely. Thus, the estimation curve provides very good feedback to a designer, and helps him perform a sound cost/performance trade-off. Once an appropriate on-chip memory size that is predicted to meet the performance requirements is selected, the final simulations can be performed and the exact configuration determined.

The results of our case study are in accordance with another study reported in [SL97] on the cache behavior of the MPEG-2 decoder. The MPEG-2 is an extension of the original MPEG algorithm. They report that the performance can possibly be improved with a *Scratch memory* by selectively storing critical data on-chip. This notion is exactly the same as the Scratch-Pad memory concept we have presented. Our technique and experimental results show that the performance is indeed enhanced by exploiting this architectural innovation. It should be noted that the cache performance for larger caches is different in our study than that in [SL97] because the original SpecCharts model from which we extracted the memory references, used a single picture block instead of the entire picture.

7.4 SUMMARY

We presented a detailed case study of the MPEG decoder to evaluate the efficacy of our analytical memory exploration technique. Experiments show that our estimation provides a very good approximation to the actual memory performance. Thus, the memory exploration can provide very useful feedback to a system designer in the early phases of system design. The estimation capability could be improved by incorporating the effects of data reuse across multiple loops.

8 CONCLUSIONS

The design of embedded Systems-on-Chip (SOC) is influenced by several evolutionary trends, such as the increase in design complexity made possible through shrinking features sizes coupled with larger die sizes for chips. On the other hand, the increasing complexity of building blocks leads to a revolutionary challenge in the emerging system design process that combines complex Intellectual-Property (IP) library blocks to create specialized embedded SOC under tight time-to-market deadlines. Such IP libraries frequently consist of pre-designed mega-cells such as processor cores and memories; several new architectural and optimization issues arise in the process of incorporation of these complex components in application-specific embedded systems. In this book, we described techniques for incorporating memory architectures in such embedded systems that contain both custom-synthesized hardware, as well as programmable processor cores.

For custom-synthesized hardware, we first showed how the organization of modern DRAMs can be exploited to yield significant performance improvements during the automatic synthesis of behavioral-level descriptions. We presented models for representation of memory accesses, and transformation techniques for incorporating page mode, and other efficient access features during synthesis. Page mode accesses can also be exploited to effect power optimizations during memory accesses. We then presented a strategy for minimizing memory access related power by suitably mapping behavioral arrays to memory.

For the programmable subsystem in embedded SOC, we presented several memory-related optimizations that result in improved system performance. First, we addressed data organization for improved performance. Storage of behavioral scalar and array variables in memory directly impacts the performance of behaviors that execute on embedded processors. Due to the relatively longer compilation time available for embedded processor-based systems, it is possible to perform a more detailed analysis of the behavior. We outlined techniques for clustering scalar variables and assigning addresses to scalar and array variables for enhancing the cache performance. Next, we described a data alignment technique for improved performance that combines loop blocking (through the selection of an appropriate tile size) together with padding.

We then addressed architectural modifications of an embedded SOC's memory subsystem. The application-specific nature of embedded systems permit the incorporation of relatively unconventional components, compared to the general purpose processing domain. One such component is Scratch-Pad memory – on-chip memory to which the assignment of data is compiler controlled. We showed how the array sizes, access frequency, and other characteristics can be factored into a data partitioning strategy between on-chip Scratch-Pad SRAM and off-chip DRAM.

Finally, we outlined a memory exploration strategy for embedded SOC. An embedded processor-based system allows for customization of on-chip memory configuration according to the requirements of the specific application. Since individual applications have widely varying memory characteristics, it is essential to generate an efficient memory organization for a given application. We presented a fast memory exploration strategy that combines the memory assignment and SRAM/DRAM data partitioning techniques with a cache performance estimation scheme to evaluate the expected performance of an application for different on-chip memory architectures. The speed advantage of such an analytical approach makes it an attractive tool for early system design exploration.

8.1 EXTENDING THIS WORK

This book described several techniques for memory organization, optimization, and exploration of embedded SOC. Each technique was focused on a single application, and exploited locality of data accesses with the goal of improving the performance of the embedded SOC. Many of these techniques also concurrently reduce power (primarily through reduced off-chip accesses). Code size (or density) is another important design constraint for embedded systems. There are a number of possible extensions arising from the work presented in this book. We first outline a few direct extensions from the material in various chapters, and then outline extensions in the area of power reduction.

Newer DRAM families

Memory optimization possibilities for DRAMs will continue to evolve in future generations of memory technology. For instance, newer DRAM families such as Synchronous DRAM (SDRAM), encompass all the efficient modes described in Chapter 3, but also have additional architectural innovations, such as dual-bank storage, which opens up the possibility for more performance and power optimization techniques. Similarly, RAMBUS DRAM, which is likely to become a popular memory device in the future, has a more complex protocol that would require an advanced modeling strategy. The scenario of an off-chip DRAM augmented by on-chip cache forms an interesting enhancement to our proposed memory access model for custom-synthesized hardware in the embedded SOC.

Memory Subsystem Exploration

In the memory exploration phase, we considered the cache line size and cache size parameters. However, other cache features such as write policy, associativity, and replacement policy could also be incorporated to expand the exploration space. As the example in Section 6.2 illustrates, the best cache size for an application is different, depending on whether a write-through or write-back policy is adopted. Further, we have assumed in this work that the access delay for the data cache and Scratch-Pad memory does not change with the size. This assumption is valid as long as the delay is less than the cycle time of the processor core. Although this handles a large number of memory configurations, for larger memory sizes that result in long delays (e.g., with on-chip DRAM), the on-chip memory access time would lie on the critical path, and the performance estimate should incorporate a more accurate cost model.

Extensions to Instruction Caches

If the code executed in an embedded processor-based system is small enough, it can be stored in on-chip ROM. However, if the code is large, an instruction cache has to be used. In this case, interesting issues arise with respect to the sizing of the instruction cache. The memory exploration strategy should, ideally, also consider the impact of instruction cache sizing on the code performance. For example, memory size trade-off studies for a given total amount of on-chip memory should include the relative sizes of on-chip data and instruction memory in addition to Scratch-Pad memory to find an ideal balance that maximizes performance.

Memory Optimizations for Power Reduction

In Chapter 3, we discussed methods for organizing data in memory to achieve power reduction in synthesized behaviors. Memory accesses also play an important role in the portion of an embedded system mapped into software [TMW94, BdM98].

The memory optimizations discussed in the embedded processor-based scenario of Chapters 4, 5, 6, and 7 have a direct impact on system-level power consumption.

Data Organization and Alignment: These optimizations focussed on the minimization of cache misses, leading to reduced cycle count for program execution. However, the objective of reducing cache misses, which reduces off-chip memory accesses, also achieves power reduction, since reduction in number of accesses is the primary source of power reduction in memories.

Scratch-Pad Memory: Here, a small on-chip Scratch-Pad memory was used to possibly reduce a large number of cache misses by storing frequently accessed data on-chip. Since on-chip memory accesses (to either on-chip cache or Scratch-Pad memory) incur far less power consumption than off-chip accesses, this optimization also helps reduce power dissipation.

Memory Exploration: Here, we generated a graph of the estimated performance for different on-chip memory sizes so that a system designer could select an efficient memory organization for the given application. To generate the corresponding graph for power dissipation during memory accesses, we can use an almost identical formulation, where the cache miss penalty (the number of processor cycles required for accessing a cache line from off-chip memory – $K + L$ in Section 6.4) is now replaced by K', the ratio of power dissipation due to off-chip and on-chip memory accesses. The value of $K + L$ is 10-20 in modern processors (Chapter 4), while that of K' is in the region of 500 (Chapter 3).

8.2 FUTURE DIRECTIONS

In the future, we can expect to see embedded SOC that contain large amounts of DRAM merged with processors and logic, as well as multiple processors on a single chip. This opens up the possibility of many novel on-chip architectures, including combinations of on-chip vector processing and on-chip symmetric multi-processing (SMP). Furthermore, several issues arise from the complex interactions between the (real-time operating) system software, instructions, data, co-processors and custom hardware for such embedded SOC. Many of these embedded SOC will be designed for memory-intensive applications; thus there will be a critical need for system-level exploration environments that allow system designers to evaluate memory configurations and partitionings based on different types of input descriptions.

The system designer is thus faced with the additional challenges of dealing with multi-processors on a chip, with the attendant tasks of coupling techniques and ideas from parallel processing, computer architecture, compilers, CAD and operating systems. Software will increasingly become the dominant IP for future embedded SOC, and will thus present major challenges for rapid design and verification. Thus designers of future embedded SOC need a strong educational grounding in each of these areas, and this has some important and serious ramifications on the design process of future embedded systems.

References

[AC91] I. Ahmad and C. Y. R. Chen. "Post-processor for data path synthesis using multiport memories,". In *Proceedings of the IEEE International Conference on Computer Aided Design*, pages 276–279, November 1991.

[ACG+95] K. Au, P. Chang, C. Giles, E. Hadad, R. Huang, D. Jones, E. Kristenson, M. Kwong, D. Lin, M. Murzello, B. Singh, S. Wan B. Wang, and F. Worrel. "MiniRISC(tm) CW4001 – a small, low-power MIPS CPU core,". In *Proceedings, Custom Integrated Circuits Conference*, May 1995.

[AP93] A. Agarwal and S. D. Pudar. "Column-associative caches: a technique for reducing the miss rate of direct-mapped caches,". In *20th International Symposium on Computer Architecture*, pages 276–279, May 1993.

[ASU93] A. V. Aho, R. Sethi, and J. D. Ullman. *Compilers: Principles, Techniques and Tools*. Addison-Wesley, 1993.

[Aus96] T. M. Austin. *Hardware and Software Mechanisms for Reducing Load Latency*. PhD thesis, University of Wisconsin – Madison, April 1996.

[Bak88] H.B. Bakoglu. *Circuits, Interconnections, and Packaging for VLSI*. Addison-Wesley, Reading, MA, 1988.

[BCM94] F. Balasa, F. Catthoor, and H. D. Man. "Dataflow-driven memory allocation for multi-dimensional signal processing systems,". In *Proceedings of the IEEE/ACM International Conference on Computer Aided Design*, November 1994.

[BCM95] F. Balasa, F. Catthoor, and H. D. Man. "Background memory area estimation for multidimensional signal processing systems,". *IEEE Transactions on VLSI Systems*, 3(2):157–172, June 1995.

[BdM98] L. Benini and G. de Micheli. *DYNAMIC POWER MANAGEMENT: Design Techniques and CAD Tools*. Kluwer Academic Publishers, norwell, MA, 1998.

[BdMM+98] L. Benini, G. de Micheli, E. Macii, D. Sciuto, and C. Silvano. "Address bus encoding techniques for system-level power optimization,". In *Design, Automation and Test in Europe*, Paris, France, February 1998.

[BENP93] U. Banerjee, R. Eigenmann, A. Nicolau, and D.A. Padua. "Automatic program parallelization,". *Proceedings of the IEEE*, 81(2), February 1993.

[BG95] S. Bakshi and D. Gajski. "A memory selection algorithm for high-performance pipelines,". In *Proceedings of the European Design Automation Conference*, 1995.

[BGS94] D. F. Bacon, S. L. Graham, and Oliver J. Sharp. "Compiler transformations for high-performance computing,". *ACM Computing Surveys*, 26(4), December 1994.

[BMB+88] M. Balakrishnan, A. K. Majumdar, D. K. Banerji, J. G. Linders, and J. C. Majithia. "Allocation of multiport memories in data path synthesis,". *IEEE Transactions on Computer-Aided Design of Integrated Circuits and Systems*, 7(4):536–540, April 1988.

[BMMZ95] P. Baglietto, M. Maresca, M. Migliardi, and N. Zingirian. "Image processing on high performance RISC systems,". Technical Report TR SM-IMP/DIST/08, University of Genoa, December 1995.

[BRSV87] R. K. Brayton, R. Rudell, and A. Sangiovanni-Vincentelli. "MIS: A multiple level logic optimization system,". *IEEE Transactions on Computer-Aided Design of Integrated Circuits and Systems*, CAD-6(6):1062–1082, November 1987.

[BS97] M. Breternitz and R. Smith. "Enhanced compression techniques to simplify program decompression and execution,". In *IEEE International Conference on Computer Design: VLSI in Computers & Processors*, pages 170–176, October 1997.

[CAC+81] G. Chaitin, M. Auslander, A. Chandra, J. Coocke, M. Hopkins, and P. Markstein. "Register allocation via coloring,". *Computer Languages*, 6, January 1981.

[CDN97] A. Capitanio, N. D. Dutt, and A. Nicolau. "Partitioning of variables for multiple-register-file architectures via hypergraph coloring,". *IFIP Transactions: Parallel Architectures and Compilation Techniques*, A-50:319–322, 1997.

[CGL96] K.-S. Chung, R. K. Gupta, and C. L. Liu. "Interface co-synthesis techniques for embedded systems,". In *Proceedings of the IEEE/ACM International Conference on Computer Aided Design*, pages 442–447, November 1996.

[CKP91] D. Callahan, K. Kennedy, and A. Porterfield. "Software prefetching,". In *International Conference on Architectural Support for Programming Languages and Operating Systems*, pages 40–52, Santa Clara, CA, April 1991.

[CL91] E. G. Coffman, Jr. and G. S. Lueker. *Probabilistic Analysis of Packing and Partitioning Algorithms*. Wiley-Interscience, New York, 1991.

[CL95] M. Cierniak and W. Li. "Unifying data and control transformations for distributed shared memory machines,". In *ACM SIGPLAN'95 Conference on Programming Language Design and Implementation*, pages 205–217, La Jolla, CA, June 1995.

[CLR92] T. H. Cormen, C. E. Leiserson, and R. L. Rivest. *Introduction to Algorithms*. The MIT Press, Cambridge, MA, 1992.

[CM95] S. Coleman and K. S. McKinley. "Tile size selection using cache organization and data layout,". In *ACM SIGPLAN'95 Conference on Programming Language Design and Implementation*, pages 279–289, La Jolla, CA, June 1995.

[CMCH91] W. Y. Chen, S. A. Mahlke, P. P. Chang, and W. W. Hwu. "Data access microarchitectures for superscalar processors with compiler-assisted data prefetching,". In *Proceedings of Microcomputing*, 1991.

[COB95] P. Chou, R. Ortega, and G. Borriello. "Interface co-synthesis techniques for embedded systems,". In *Proceedings of the IEEE/ACM International Conference on Computer Aided Design*, pages 280–287, November 1995.

[CS93] F. Catthoor and L. Svensson. *Application-Driven Architecture Synthesis*. Kluwer Academic Publishers, 1993.

[EK91] P. M. Embree and B. Kimble. *C Language Algorithms for Digital Signal Processing*. Prentice Hall, Englewood Cliffs, 1991.

[Ess93] K. Esseghir. “Improving data locality for caches,”. Master’s thesis, Department of Computer Science, Rice University, 1993.

[Fly95] M. J. Flynn. *Computer architecture – pipelined and parallel processor design*. Jones and Bartlett Publishers, 1995.

[Gal91] D. Le Gall. “MPEG: A video compression standard for multimedia applications,”. *Communications of the ACM*, 34(4):46–58, April 1991.

[GCM98] E. De Greef, F. Catthoor, and H. De Man. “Memory size reduction through storage order optimization for embedded parallel multimedia applications,”. In *International Symposium on Parallel Processing*, pages 125–129, Orlando, FL, April 1998.

[GDLW92] D. Gajski, N. Dutt, S. Lin, and A. Wu. *High Level Synthesis: Introduction to Chip and System Design*. Kluwer Academic Publishers, 1992.

[Geb97] C. Gebotys. “Dsp address optimization using a minimum cost circulation technique,”. In *Proceedings of the IEEE/ACM International Conference on Computer Aided Design*, pages 100–103, November 1997.

[GJ79] M. R. Garey and D. S. Johnson. *Computers and Intractibility – A Guide to the Theory of NP-Completeness*. W.H. Freeman, 1979.

[Gup95] R. K. Gupta. *Co-Synthesis of Hardware and Software for Digital Embedded Systems*. Kluwer Academic Publishers, Boston, 1995.

[GVNG94] D. Gajski, F. Vahid, S. Narayan, and J. Gong. *Specification and Design of Embedded Systems*. Prentice Hall, 1994.

[GW87] R. C. Gonzalez and P. Wintz. *Digital Image Processing*. Addison-Wesley, 1987.

[Hil88] M. D. Hill. “The case for direct-mapped caches,”. *IEEE Computer*, 21(12):25–41, December 1988.

[HP94] J. L. Hennessy and D. A. Patterson. *Computer Architecture – A Quantitative Approach*. Morgan Kaufman, San Francisco, CA, 1994.

[IBM96] IBM Corporation. *http://www.chips.ibm.com/products/memory/50H7629/*, 1996.

[JD97] P. K. Jha and N. Dutt. “Library mapping for memories,”. In *European Design and Test Conference*, pages 288–292, March 1997.

[Jou90] N. P. Jouppi. “Improving direct-mapped cache performance by the addition of a small fully-associative cache and prefetch buffers,”. In *International Symposium on Computer Architecture*, pages 364–373, Seattle, WA, May 1990.

[Jou93] N. P. Jouppi. “Cache write policies and performance,”. In *International Symposium on Computer Architecture*, pages 191–201, San Diego, CA, May 1993.

[KCM98] C. Kulkarni, F. Catthoor, and H. De Man. “Code transformations for low power caching in embedded multimedia processors,”. In *International Symposium on Parallel Processing*, pages 125–129, Orlando, FL, April 1998.

[KL91] A. C. Klaiber and H. M. Levy. “Architecture for software-controlled data prefetching,”. In *Proceedings of the 8th Annual International Symposium on Computer Architecture*, pages 81–85, May 1991.

[KL93] T. Kim and C. L. Liu. “Utilization of multiport memories in data path synthesis,”. In *Design Automation Conference*, pages 298–302, June 1993.

[KLPMS97] D. Kirovski, C. Lee, M. Potkonjak, and W. Mangione-Smith. “Application-driven synthesis of core-based systems,”. In *Proceedings of the IEEE/ACM International Conference on Computer Aided Design*, pages 104–107, San Jose, CA, November 1997.

[KND94] D. J. Kolson, A. Nicolau, and N. D. Dutt. “Integrating program transformations in the memory-based synthesis of image and video algorithms,”. In *Proceedings of the IEEE/ACM International Conference on Computer Aided Design*, pages 27–30, San Jose, CA, November 1994.

[KNDK96] D. J. Kolson, A. Nicolau, N. D. Dutt, and K. Kennedy. “Optimal register assignment to loops for embedded code generation,”. *ACM Transactions on Design Automation of Electronic Systems*, 1(2):251–279, April 1996.

[Koh78] Z. Kohavi. *Switching and Finite Automata Theory*. McGraw-Hill, New York, 1978.

[KP87] F. J. Kurdahi and A. C. Parker. "Real: A program for register allocation,". In *Design Automation Conference*, 1987.

[KR94] D. Karchmer and J. Rose. "Definition and solution of the memory packing problem for field-programmable systems,". In *Proceedings of the IEEE/ACM International Conference on Computer Aided Design*, pages 20–26, November 1994.

[LDK+95] S. Liao, S. Devadas, K. Keutzer, S. Tjiang, and A. Wang. "Storage assignment to decrease code size,". In *Proceedings of the ACM SIGPLAN'95 Conference on Programming Language Design and Implementation*, pages 186–195, La Jolla, CA, June 1995.

[LH98] Y. Li and J. Henkel. "A framework for estimating and minimizing energy dissipation of embedded hw/sw systems,". In *Design Automation Conference*, pages 188–193, San Francisco, CA, June 1998.

[LKMM95] T. Ly, D. Knapp, R. Miller, and D. MacMillen. "Scheduling using behavioral templates,". In *ACM/IEEE Design Automation Conference*, pages 599–604, June 1995.

[LM96] R. Leupers and P. Marwedel. "Algorithms for address assignment in dsp code generation,". In *IEEE/ACM International Conference on Computer Aided Design*, pages 109–112, San Jose, CA, November 1996.

[LMP94] C. Liem, T. May, and P. Paulin. "Instruction-set matching and selection for DSP and ASIP code generation,". In *Proceedings, European Design and Test Conference*, pages 31–37, Paris, France, February 1994.

[LMW95] Y-T. S. Li, S. Malik, and A. Wolfe. "Performance estimation of embedded software with instruction cache modeling,". In *IEEE/ACM International Conference on Computer Aided Design*, pages 380–387, San Jose, CA, November 1995.

[LPJ96] C. Liem, P. Paulin, and Ahmed Jerraya. "Address calculation for retargetable compilation and exploration of instruction-set architectures,". In *Design Automation Conference*, pages 597–600, Las Vegas, U.S.A., June 1996.

[LRW91] M. Lam, E. Rothberg, and M. E. Wolf. "The cache performance and optimizations of blocked algorithms,". In *Proceedings of the Fourth*

International Conference on Architectural Support for Programming Languages and Operating Systems, pages 63–74, April 1991.

[LSI92] LSI Logic Corporation, Milpitas. *CW33000 MIPS Embedded Processor User's Manual*, 1992.

[LW94] A. R. Lebeck and D. A. Wood. "Cache profiling and the spec benchmarks: A case study,". *IEEE Computer*, 27(10), October 1994.

[LW97] Y. Li and W. Wolf. "A task-level hierarchical memory model for system synthesis of multiprocessors,". In *Design Automation Conference*, June 1997.

[LW98] H. Lekatsas and W. Wolf. "Code compression for embedded systems,". In *Design Automation Conference*, pages 516–521, San Francisco, June 1998.

[Mar97] B. Margolin. "Embedded systems to benefit from advances in DRAM technology,". *Computer Design*, pages 76–86, March 1997.

[McF89] S. McFarling. "Program optimization for instruction caches,". In *Third International Conference on Architectural Support for Programming Languages and Operating Systems*, pages 183–191, Boston, MA, April 1989.

[MG95] P. Marwedel and G. Goosens, editors. *Code Generation for Embedded Processors*. Kluwer Academic Publishers, Norwell, MA, 1995.

[MLC97] E. Musoll, T. Lang, and J. Cortadella. "Exploiting the locality of memory references to reduce the address bus energy,". In *International Symposium on Low Power Electronics and Design*, pages 202–207, Monterey, CA, August 1997.

[MLG92] T. C. Mowry, M. S. Lam, and A. Gupta. "Design and evaluation of a compiler algorithm for prefetching,". In *Proceedings of the Fifth International Conference on Architectural Support for Programming Languages and Operating Systems*, pages 202–207, Monterey, CA, October 1992.

[Muc97] S. Muchnick. *Advanced Compiler Design and Implementation*. Morgan Kaufman, San Francisco, CA, 1997.

[PD95a] P. R. Panda and N. D. Dutt. "1995 High level synthesis design repository,". In *International Symposium on System Synthesis*, pages 170–174, Cannes, France, September 1995.

[PD95b] P. R. Panda and N. D. Dutt. "Low power memory mapping through reducing address bus activity,". Technical Report 95-32, University of California, Irvine, November 1995.

[PD96] P. R. Panda and N. D. Dutt. "Reducing address bus transitions for low power memory mapping,". In *Proceedings, European Design and Test Conference*, pages 63–67, Paris, France, March 1996.

[PD97] P. R. Panda and N. D. Dutt. "Behavioral array mapping into multiport memories targeting low power,". In *Proceedings of the 10th International Conference on VLSI Design*, pages 268–272, Hyderabad, India, January 1997.

[PDN96] P. R. Panda, N. D. Dutt, and A. Nicolau. "SRAM vs. data cache: The memory data partitioning problem in embedded systems,". Technical Report 96-42, University of California, Irvine, September 1996.

[PDN97a] P. R. Panda, , N. D. Dutt, and A. Nicolau. "Architectural exploration and optimization of local memory in embedded systems,". In *10th International Symposium on System Synthesis*, pages 90–97, Antwerp, Belgium, September 1997.

[PDN97b] P. R. Panda, N. D. Dutt, and A. Nicolau. "Data cache sizing for embedded processor applications,". Technical Report ICS-TR-97-30, University of California, Irvine, June 1997.

[PDN97c] P. R. Panda, N. D. Dutt, and A. Nicolau. "Efficient utilization of Scratch-Pad memory in embedded processor applications,". In *European Design and Test Conference*, March 1997.

[PDN97d] P. R. Panda, N. D. Dutt, and A. Nicolau. "Exploiting off-chip memory access modes in high-level synthesis,". In *Proceedings of the IEEE/ACM International Conference on Computer Aided Design*, pages 333–340, San Jose, CA, November 1997.

[PDN97e] P. R. Panda, N. D. Dutt, and A. Nicolau. "Memory data organization for improved cache performance in embedded processor applications,". *ACM Transactions on Design Automation of Electronic Systems*, 2(4), October 1997.

[PDN98a] P. R. Panda, N. D. Dutt, and A. Nicolau. "Data cache sizing for embedded processor applications,". In *Design, Automation and Test in Europe*, pages 925–926, Paris, France, February 1998.

[PDN98b] P. R. Panda, N. D. Dutt, and A. Nicolau. "Incorporating dram access modes into high-level synthesis,". *IEEE Transactions on Computer Aided Design*, 17(2):96–109, February 1998.

[PH94] D. A. Patterson and J. L. Hennessy. *Computer Organization & Design – The Hardware/Software Interface*. Morgan Kaufman, San Mateo, CA, 1994.

[PNDN97] P. R. Panda, H. Nakamura, N. D. Dutt, and A. Nicolau. "A data alignment technique for improving cache performance,". In *Proceedings IEEE International Conference on Computer Design*, pages 587–592, October 1997.

[PTVF92] W. H. Press, S. A. Teukolsky, W. T. Vetterling, and B. P. Flannery. *Numerical Recipes in C: The Art of Scientific Computing*. Cambridge University Press, 1992.

[Raw93] J. Rawat. "Static analysis of cache performance for real-time programming,". Master's thesis, Iowa State University, May 1993.

[SB95] M. R. Stan and W. P. Burleson. "Bus-invert coding for low power I/O,". *IEEE Transactions on VLSI Systems*, 3(1):49–58, March 1995.

[SL97] P. Soderquist and M. Leeser. "Memory traffic and data cache behavior of an MPEG-2 software decoder,". In *Proceedings IEEE International Conference on Computer Design*, pages 417–422, October 1997.

[SSV96] K. Suzuki and A. Sangiovani-Vincentelli. "Efficient software performance estimation methods for hardware/software codesign,". In *Design Automation Conference*, pages 605–610, June 1996.

[Sun93] Sun Microsystems Laboratories Inc., Mountain View, CA. *Shade User's Manual*, 1993.

[SWCdJ97] P. Slock, S. Wuytack, F. Catthoor, and G. de Jong. "Fast and extensive system-level memory exploration for atm applications,". In *International Symposium on System Synthesis*, pages 74–81, September 1997.

[SYO+97] B. Shackleford, M. Yasuda, E. Okushi, H. Koizumi, H. Tomiyama, and H. Yasuura. "Memory-cpu size optimization for embedded system designs,". In *Design Automation Conference*, June 1997.

[Tho95] A. Thordarson. "Comparison of manual and automatic behavioral synthesis of MPEG algorithm,". Master's thesis, University of California, Irvine, 1995.

[TIIY98] H. Tomiyama, T. Ishihara, A. Inoue, and H. Yasuura. "Instruction scheduling for power reduction in processor-based system design,". In *Design, Automation, and Test in Europe*, Paris, France, February 1998.

[TMW94] V. Tiwari, S. Malik, and A. Wolfe. "Power analysis of embedded software: a first step towards software power minimization,". *IEEE Transactions on VLSI Systems*, 2(4):437–445, December 1994.

[TS86] C. Tseng and D. P. Siewiorek. "Automated synthesis of datapaths in digital systems,". *IEEE Transactions on Computer Aided Design*, 6(7):379–395, July 1986.

[TY96a] H. Tomiyama and H. Yasuura. "Optimal code placement of embedded software for instruction caches,". In *European Design and Test Conference*, pages 96–101, Paris, France, March 1996.

[TY96b] H. Tomiyama and H. Yasuura. "Size-constrained code placement for cache miss rate reduction,". In *International Symposium on System Synthesis*, pages 96–101, La Jolla, CA, November 1996.

[TY97] H. Tomiyama and H. Yasuura. "Code placement techniques for cache miss rate reduction,". *ACM Transactions on Design Automation of Electronic Systems*, 2(4), October 1997.

[VSR94] I. M. Verbauwhede, C. J. Scheers, and J. M Rabaey. "Memory estimation for high level synthesis,". In *Design Automation Conference*, June 1994.

[Wal91] G. K. Wallace. "The JPEG still picture compression standard,". *Communications of the ACM*, 34(4):30–44, April 1991.

[WCF+94] S. Wuytack, F. Catthoor, F. Franssen, L. Nachtergaele, and H.De Man. "Global communication and memory optimizing transformations for low power systems,". In *IEEE International Workshop on Low Power Design*, pages 203–208, April 1994.

[WCM95] S. Wuytack, F. Catthoor, and H.De Man. "Transforming set data types to power optimal data structures,". In *International Symposium on Low Power Design*, pages 51–56, April 1995.

[WE85] N. H. E. Weste and K. Eshraghian. *Principles of CMOS VLSI design : a systems perspective*. Addison-Wesley, 1985.

[Wil97] R. Wilson. "Graphics IC vendors take a shot at embedded DRAM,". *Electronic Engineering Times*, (938):41,57, January 27 1997.

[WL91] M. E. Wolf and M. Lam. "A data locality optimizing algorithm,". In *Proceedings of the SIGPLAN'91 Conference on Programming Language Design and Implementation*, pages 30–44, Toronto, Canada, June 1991.

[Wol89] M. J. Wolfe. "More iteration space tiling,". In *Proceedings of Supercomputing*, pages 655–664, Reno, NV, November 1989.

[YJH+95] Y. Yamada, T. L. Johnson, G. Haab, J. C. Gyllenhaal, and W. W. Hwu. "Reducing cache misses in numerical applications using data relocation and prefetching,". Technical Report CRHC-95-04, University of Illinois, Urbana, 1995.

Index